THE LIFE OF

SIR RICHARD BURTON

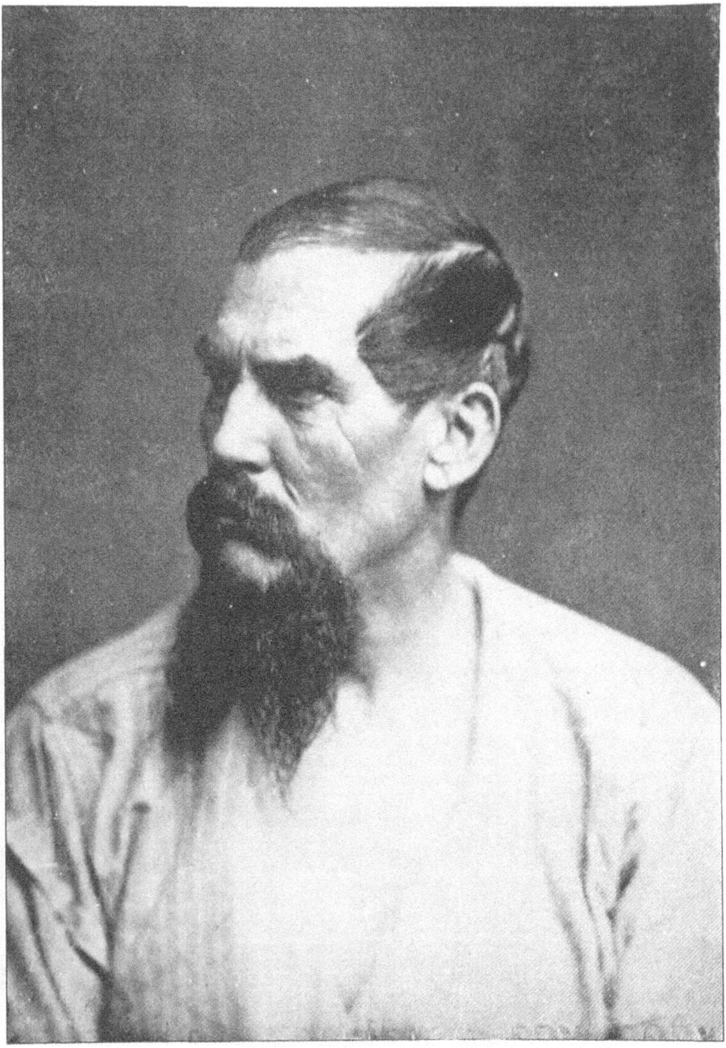

SIR RICHARD BURTON

From a photograph by Elliott & Fry

THE LIFE OF
SIR RICHARD BURTON

BY

THOMAS WRIGHT
AUTHOR OF "THE LIFE OF EDWARD FITZGERALD," ETC.

WITH SIXTY-FOUR PLATES.

TWO VOLUMES
VOL. I

LONDON
EVERETT & CO.
1906

This Work is
Dedicated to
Sir Richard Burton's Kinsman
and Friend,
Major St. George Richard Burton,
The Black Watch.

PREFACE.

Fifteen years have elapsed since the death of Sir Richard Burton and twelve since the appearance of the biography by Lady Burton. A deeply pathetic interest attaches itself to that book. Lady Burton was stricken down with an incurable disease. Death with its icy breath hung over her as her pen flew along the paper, and the questions constantly on her lips were "Shall I live to complete my task? Shall I live to tell the world how great and noble a man my husband was, and to refute the calumnies that his enemies have so industriously circulated?" She did complete it in a sense, for the work duly appeared; but no one recognised more clearly than herself its numerous shortcomings. Indeed, it is little better than a huge scrap-book filled with newspaper cuttings and citations from Sir Richard's and other books, hurriedly selected and even more hurriedly pieced together. It gives the impressions of Lady Burton alone, for those of Sir Richard's friends are ignored—so we see Burton from only one point of view. Amazing to say, it does not contain a single original anecdote [1]—though perhaps, more amusing anecdotes could be told of Burton than of any other modern Englishman. It will be my duty to rectify Lady Burton's mistakes and mis-statements and to fill up the vast hiatuses that she has left. Although it will be necessary to subject her to criticism, I shall endeavour at the same time to keep constantly in mind the queenliness and beauty of her character, her almost unexampled devotion to her husband, and her anxiety that everyone should think well of him. Her faults were all of the head. Of the heart she had absolutely none.

As the Richard Burton whom I have to pourtray

[1] The few anecdotes that Lady Burton does give are taken from the books of Alfred B. Richards and others.

differs considerably from Lady Burton's "Earthly God,"[1] I have been very careful to give chapter and verse for all my statements. The work has been written on the same lines as my *Life of Edward FitzGerald*; that is to say, without any aim except to arrive at the precise truth. But although I have regarded it as no concern of mine whether any particular fact tells for or against Sir Richard Burton, I do think that when the reader rises from the last page he will feel that he has been in the company not only of one of the greatest, noblest and most fearless of Englishmen, but also of one who, without making much profession of doing so, really loved his fellow-men, and who, despite his inability to put himself in line with religionists, fought steadily on the side of righteousness. We are aware that there are in his books a few observations which call for vehement and unqualified denunciation; but against them must be placed the fundamental goodness of the man, to which all who knew him intimately have testified. In not a few respects Sir Richard Burton's character resembled Edward FitzGerald's. Burton, indeed, hailed the adapter of *Omar Khayyam* as a "fellow Sufi."

Lady Burton, too, comes extremely well out of the fire of criticism. The reader may object to her religious views, he may smile at her weaknesses, he may lament her indiscretions, but he will recognise that at bottom she was a God-fearing, noble-minded woman; and he will, we think, find himself really in love with her almost before knowing it.

The amount of absolutely new information in this work is very large. Thus we are telling for the first time the history of Burton's friendships with Mr. F. F. Arbuthnot, Mr. John Payne, and others; and we are giving for the first time, too, a complete and accurate history of the translation of *The Arabian Nights*, *The Scented Garden*, and other works. Hundreds of new facts are recorded

(1) Lady Burton to Mrs. E. J. Burton, 23rd March 1891. See Chapter xxxix.

respecting these and other absorbing topics, while the citations from the unpublished letters of Burton and Lady Burton will, we are sure, receive a welcome. We are able to give about fifty entirely new anecdotes—many of them extremely piquant and amusing. We also tell the touching story of Burton's brother Edward. In our accounts of Burton's travels will be found a number of interesting facts and some anecdotes not given in Burton's works.

The new material has been derived from many sources—but from ten in particular.

(1) From two hundred unpublished letters of Sir Richard Burton and Lady Burton.

(2) From interviews with Mrs. E. J. Burton[1] and Mr. F. Burton (Burton's cousins), Mr. John Payne, Mrs. Arbuthnot, Mr. Watts-Dunton, Mr. W. F. Kirby, Mr. A. G. Ellis, Dr. Codrington, Professor James F. Blumhardt, Mr. Henry R. Tedder (librarian and secretary of The Athenæum, Burton's club), Mrs. Baddeley (mother of Burton's friend, St. Clair Baddeley), Madame Nicastro (sister of the late Mr. Albert Letchford, illustrator of *The Arabian Nights*), Dr. Grenfell Baker (Burton's medical attendant during the last three years of his life), and many other ladies and gentlemen.

(3) From letters received from Major St. George Burton (to whom I have the pleasure of dedicating this work), Lady Bancroft, Mr. D. MacRitchie, Mr. E. S. Mostyn Pryce (representative of Miss Stisted), Gunley Hall, Staffordshire, M. Charles Carrington, of Paris, who sent me various notes, including an account of Burton's unfinished translation of Apuleius's *Golden Ass*, the MS. of which is in his possession, the Very Rev. J. P. Canon McCarthy, of Ilkeston, for particulars of "The Shrine of our Lady of Dale," Mr. Segrave (son of Burton's

[1] A three days' visit to Brighton, where I was the guest of Mrs. E. J. Burton, is one of the pleasantest of my recollections.

"dear Louisa"), Mrs. Agg (Burton's cousin), and Mr. P. P. Cautley (Burton's colleague at Trieste). Nor must I omit reference to a kind letter received from Mrs. Van Zeller, Lady Burton's only surviving sister.[1]

(4) From the Burton collections in the Free Libraries of Camberwell and Kensington.

(5) From unpublished manuscripts written by Burton's friends.

(6) From the church registers at Elstree. By examination of these and other documents I have been able to correct many mistakes.

(7) From the manuscripts of F. F. Arbuthnot and the Oriental scholar, Edward Rehatsek. These are now in the possession of the Royal Asiatic Society.

(8) From Mr. Arbuthnot's typewritten and unpublished *Life of Balzac* now in my possession. This contains many notes throwing light on the Burton and Arbuthnot friendship.

(9) From the Genealogical Table of the Burtons of Shap, very kindly sent me by Mr. E. S. Mostyn Pryce

(10) From various persons interviewed during many journeys. One of these journeys (June 1905) took me, of course, to the Tomb at Mortlake, and I was gratified to find that, owing to the watchfulness of the Arundell family, it is kept in perfect repair.[2]

Let me first speak of the unpublished letters. These were lent me by Mr. John Payne (40 letters), Mr. W. F. Kirby (50 letters), Major St. George Burton, Mrs. E. J. Burton, Mrs. Agg, Mr. Mostyn Pryce, Dr. Tuckey, Mr. D. MacRitchie, and Mr. A. G. Ellis. Many of the letters reveal Burton in quite a new light. His patriotism and

(1) Mrs. Van Zeller had, in the first instance, been written to, in my behalf, by Mrs. E. J. Burton.

(2) It is important to mention this because a few months ago a report went the round of the newspapers to the effect that the tomb was in ruins.

his courage were known of all men, but the womanly tenderness of his nature and his intense love for his friends will come to many as a surprise. His distress, for example, on hearing of the death of Drake, [1] is particularly affecting.

Of the friends of Sir Richard Burton who have been interviewed I must mention first of all Mr. John Payne. But for Mr. Payne's generous assistance, this work, I must frankly admit, could not have been written. He, and he alone, held the keys to whole chambers of mystery. Mr. Payne was at first extremely reluctant to give me the material required. Indeed, in his first letter of reply to my request for information (7th August 1904) he declined positively either to enter the lists against Burton, with whom, he said, he had been on terms of intimate friendship, or to discuss the matter at all. "As for what," he said, "it pleases the public to think (save the mark!) of the relative merits of my own and Burton's translations, I have long ceased to care a straw." But this led me to write even more pressingly. I assured Mr. Payne that the public had been unjust to him simply because nobody had hitherto set himself the great task of comparing the two translations, and because the true history of the case had never been laid before them. I assured him that I yielded to nobody in admiration of Sir Richard Burton—that is, on account of what he (Sir Richard) did do, not on account of what he did not do; and I gave it as my opinion that Mr. Payne owed it both to the public and to himself to lay bare the whole story. After several letters and interviews I at last induced him to give way; and I think the public will thank me for my persistency.

My revelations, which form an astonishing story, will no doubt come as a complete surprise to almost everybody. I can imagine them, indeed, dropping like a bombshell into some circles; but they are founded, not only upon conversations with Mr. Payne, but upon Burton's own letters to Mr. Payne, all of which have been in my

(1) See Chapter xvii.

hands, and careful study of the two translations. The public, however, cannot possibly be more surprised than I myself was when the actual facts gradually disclosed themselves. When I compared the two translations, page by page, I could scarcely believe my own eyes ; and only one conclusion was possible. Burton, indeed, has taken from Payne at least three-quarters of the entire work. He has transferred many hundreds of sentences and clauses bodily. Sometimes we come upon a whole page with only a word or two altered.[1] In short, amazing to say, the public have given Burton credit for a gift which he did not possess[2]—that of being a great translator. If the public are sorry, we are deeply sorry, too, but we cannot help it. Burton's exalted position, however, as ethnologist and anthropologist, is unassailable. He was the greatest linguist and traveller that England ever produced. And four thrones are surely enough for any man. I must mention that Mr. Payne gave me an absolutely free hand—nay, more than that, having placed all the documents before me, he said—and this he repeated again and again—"Wherever there is any doubt, give Burton the benefit of it," and I have done so.

In dealing with the fight[3] over *The Arabian Nights* I have endeavoured to write in such a way as to give offence to nobody, and for that reason have made a liberal use of asterisks. I am the more desirous of saying this because no one is better aware than myself of the services that some of Burton's most bitter opponents—those ten or twelve men whom he contemptuously termed Laneites—have rendered to literature and knowledge. In short, I regard the battle as fought and won. I am merely

(1) It is as if someone were to write "Allah is my shepherd, I shall not want," &c., &c.—here and there altering a word—and call it a new translation of the Bible.

(2) See almost any 'Cyclopædia. Of the hundreds of persons with whom I discussed the subject, one, and only one, guessed how matters actually stood—Mr. Watts-Dunton.

(3) Between Payne and Burton on the one side and the adherents of E. W. Lane on the other

writing history. No man at the present day would dream of mentioning Lane in the same breath with Payne and Burton. In restoring to Mr. Payne his own, I have had no desire to detract from Burton. Indeed, it is impossible to take from a man that which he never possessed. Burton was a very great man, Mr. Payne is a very great man, but they differ as two stars differ in glory. Burton is the magnificent man of action and the anthropologist, Mr. Payne the brilliant poet and prose writer. Mr. Payne did not go to Mecca or Tanganyika, Burton did not translate *The Arabian Nights*,(1) or write *The Rime of Redemption* and *Vigil and Vision*. He did, however, produce the annotations of *The Arabian Nights*, and a remarkable enough and distinct work they form.

I recall with great pleasure an evening spent with Mr. Watts-Dunton at The Pines, Putney. The conversation ran chiefly on the Gipsies,(2) upon whom Mr. Watts-Dunton is one of our best authorities, and the various translations of *The Arabian Nights*. Both he and Mr. A. C. Swinburne have testified to Burton's personal charm and his marvellous powers. "He was a much valued and loved friend," wrote Mr. Swinburne to me, (3) "and I have of him none but the most delightful recollections." Mr. Swinburne has kindly allowed me to give in full his magnificent poem on "The Death of Richard Burton." Dr. Grenfell Baker, whom I interviewed in London, had much to tell me respecting Sir Richard's last three years; and he has since very kindly helped me by letter.

The great object of this book is to tell the story of Burton's life, to delineate as vividly as possible his remarkable character—his magnetic personality, and to defend him alike from enemy and friend. In writing it my difficulties have been two. First, Burton himself was woefully inaccurate as an autobiographer, and we

(1) At the very outside, as before stated, only about a quarter of it can by any stretch of the imagination be called his.

(2) Burton's work on this subject will be remembered.

(3) 31st July 1905.

must also add regretfully that we have occasionally found him colouring history in order to suit his own ends.[1] He would have put his life to the touch rather than misrepresent if he thought any man would suffer thereby; but he seems to have assumed that it did not matter about keeping strictly to the truth if nobody was likely to be injured. Secondly, Lady Burton, with haughty indifference to the opinions of everyone else, always exhibited occurrences in the light in which she herself desired to see them. This fact and the extreme haste with which her book was written are sufficient to account for most of its shortcomings. She relied entirely upon her own imperfect recollections. Church registers and all such documents were ignored. She begins with the misstatement that Burton was born at Elstree, she makes scarcely any reference to his most intimate friends and even spells their names wrongly.[2] Her remarks on the *Kasídah* are stultified by the most cursory glance at that poem; while the whole of her account of the translating of *The Arabian Nights* is at variance with Burton's own letters and conversations. I am assured by several who knew Burton intimately that the untrustworthiness of the latter part of Lady Burton's "Life" of her husband is owing mainly to her over-anxiety to shield him from his enemies. But I think she mistook the situation. I do not believe Burton had any enemies to speak of at the time of his death.

If Lady Burton's treatment of her husband's unfinished works cannot be defended, on the other hand I shall show that the loss as regards *The Scented Garden* was chiefly a pecuniary one, and therefore almost entirely her own. The publication of *The Scented Garden* would not—it could not—have added to Burton's fame. However, the matter will be fully discussed in its proper place.

(1) See Chapters xxii. to xxix. and xxxv. He confessed to having inserted in *The Arabian Nights* a story that had no business there. See Chapter xxix., § 136.

(2) Thus she calls Burton's friend Da Cunha, Da Gama, and gives Arbuthnot wrong initials.

It has generally been supposed that two other difficulties must confront any conscientious biographer of Burton—the first being Burton's choice of subjects, and the second the friction between Lady Burton and the Stisteds. But as regards the first, surely we are justified in assuming that Burton's studies were pursued purely for historical and scientific purposes. He himself insisted in season and out of season that his outlook was solely that of the student, and my researches for the purposes of this work have thoroughly convinced me that, however much we may deprecate some of these studies, Burton himself was sincere enough in his pursuit of them. His nature, strange as it may seem to some ears, was a cold one;[1] and at the time he was buried in the most forbidding of his studies he was an old man racked with infirmities. Yet he toiled from morning to night, year in year out, more like a navvy than an English gentleman, with an income of £700 a year, and 10,000 "jingling, tingling, golden, minted quid," as R. L. Stevenson would have said, in his pocket. In his hunger for the fame of an author, he forgot to feed his body, and had to be constantly reminded of its needs by his medical attendant and others. And then he would wolf down his food, in order to get back quickly to his absorbing work. The study had become a monomania with him.

I do not think there is a more pathetic story in the history of literature than that which I have to tell of the last few weeks of Burton's life. You are to see the old man, always ailing, sometimes in acute pain—working twenty-five hours a day as it were—in order to get completed a work by which he supposed he was to live for ever. In the same room sits the wife who dearly loves him, and whom he dearly loves and trusts. A few days pass. He is gone. She burns, page by page, the work at which he had toiled so long and so patiently. And here comes the pathos of it—

(1) I mean in a particular respect, and upon this *all* his friends are agreed. But no man could have had a warmer heart.

she was, in the circumstances, justified in so doing. As regards Lady Burton and the Stisteds, it was natural, perhaps, that between a staunch Protestant family such as the Stisteds, and an uncompromising Catholic like Lady Burton there should have been friction; but both Lady Burton and Miss Stisted are dead. Each made, during Lady Burton's lifetime, an honest attempt to think well of the other; each wrote to the other many sweet, sincere, and womanly letters; but success did not follow. Death, however, is a very loving mother. She gently hushes her little ones to sleep; and, as they drop off, the red spot on the cheek gradually fades away, and even the tears on the pillow soon dry.

Although Miss Stisted's book has been a help to me I cannot endorse her opinion that Burton's recall from Damascus was the result of Lady Burton's indiscretions. Her book gives some very interesting reminiscences of Sir Richard's childhood and early manhood, [1] but practically it finishes with the Damascus episode. Her innocent remarks on *The Scented Garden* must have made the anthropological sides of Ashbee, Arbuthnot, and Burton's other old friends shake with uncontrollable laughter. Unfortunately, she was as careless as Lady Burton. Thus on page 48 she relates a story about Burton's attempt to carry off a nun; but readers of Burton's book on Goa will find that it had no connection with Burton whatever. It was a story someone had told him.

In these pages Burton will be seen on his travels, among his friends, among his books, fighting, writing, quarrelling, exploring, joking, flying like a squib from place to place—a 19th century Lord Peterborough, though with the world instead of a mere continent for theatre. Even late in life, when his infirmities prevented larger circuits, he careered about Europe in a Walpurgic style that makes the mind giddy to dwell upon.

(1) Particularly pretty is the incident of the families crossing the Alps, when the children get snow instead of sugar. See Chapter iv.

Of Burton's original works I have given brief summaries ; but as a writer he shines only in isolated passages. We go to him not for style but for facts. Many of his books throw welcome light on historical portions of the Bible.[1]

Of those of his works which are erotic in the true sense of the word I have given a sufficient account, and one with which I am convinced even the most captious will not find fault.[2] When necessity has obliged me to touch upon the subject to which Sir Richard devoted his last lustrum, I have been as brief as possible, and have written in a way that only scholars could understand. In short I have kept steadily in view the fact that this work is one which will lie on drawing-room tables and be within the reach of everyone. I have nowhere mentioned the subject by name, but I do not see how I could possibly have avoided all allusion to it. I have dwelt on Burton's bravery, his tenderness, his probity, his marvellous industry, his encyclopædic learning—but the picture would not have been a true one had I entirely overpassed the monomania of his last days. Hamlet must be shown, if not at his maddest, at any rate mad, or he would not be Hamlet at all.

As regards Burton's letters, I have ruthlessly struck out every sentence that might give offence.[3] While I have not hesitated to expose Sir Richard's faults, I have endeavoured to avoid laying too much stress upon them. I have tried, indeed, to get an idea of the mountain not only by climbing its sides, but also by viewing it from a distance. I trust that there will be found nothing in this book to hurt the feelings of any living person or indeed

(1) Particularly *Unexplored Syria* and his books on Midian.

(2) It will be noticed, too, that in no case have I mentioned where these books are to be found. In fact, I have taken every conceivable precaution to make this particular information useless except to *bona-fide* students.

(3) I am not referring to "Chaucerisms," for practically they do not contain any. In some two hundred letters there are three Chaucerian expressions. In these instances I have used asterisks, but, really, the words themselves would scarcely have mattered. There are as plain in the *Pilgrim's Progress*.

of any body of persons. I have certainly tried my utmost to avoid causing pain, and if the reader will kindly bear in mind that it is as much a Christian duty to avoid taking offence as to avoid giving offence, we shall amble along pleasantly together to the very last page. Out of consideration for Catholics I have suppressed a number of passages ; and if I have allowed Sir Richard in one or two instances to make a lunge at their church, I trust they will notice that I have permitted him the same licence with regard to the Church of England and Exeter Hall. Finally, my impartiality is proved by my allowing him to gird at the poet Cowper.

Wherever possible, that is to say, when I could do it without ambiguity I have also out of courtesy used the term Catholic instead of Roman Catholic ; and in order to meet what I believe to be the wishes of Lady Burton's executors, I have omitted all mention of certain events that occurred after Sir Richard's death.

The various works of Mr. W. H. Wilkins have been of great help to me, and I cannot avoid paying a passing tribute to the excellent opening passages (1) of the Preface to his edition of Lady Burton's Life of her husband.

The illustrations in this book are of exceptional interest. They include the Burton family portraits, the originals of which are in the possession of Mr. Mostyn Pryce and Mrs. Agg. During the lifetime of Sir Richard and Lady Burton they were the property of Lady and Miss Stisted ; but, owing to her difference with these ladies, Lady Burton was not able to use them in the life of her husband ; and Miss Stisted's own scheme did not include illustrations. So they are now reproduced for the first time. The most noticeable are the quaint picture of Burton, his brother and sister as children, and the oil painting of Burton and Lady Stisted made by Jacquand about 1851. Of great interest, too, is the series of photographs taken

(1) I have often thought that the passage "I often wonder. . . . given to the world to-day," contains the whole duty of the conscientious biographer in a nutshell.

at Trieste by Dr. Grenfell Baker; while the portraits of Burton's friends, Mr. F. F. Arbuthnot, Mr. John Payne, Major St. George Burton, Dr. Baker, Mr. W. F. Kirby, Mr. A. G. Ellis, Professor J. F. Blumhardt, and others, will no doubt be appreciated by the public.

The writing of this book has been a thorough pleasure to me, not only on account of the infinite charm of the subject, but also because everyone whom I have approached has treated me with studied kindness. The representatives of Sir Richard Burton, of Lady Burton (through Mr. W. H. Wilkins) and of Miss Stisted have not only helped and permitted me to use the unpublished letters,[1] but have generously given me a free hand. I am deeply indebted to them, and I can only trust that these pages will prove that their confidence in my judgment has not been misplaced.

To everyone who has assisted me I tender my sincere thanks, and I assure them that I shall never forget their abundant kindness.

Finally, in writing this work every possible care has been taken to ensure accuracy[2]; but that absolute perfection has been attained is improbable. It is hoped, however,—to borrow the quaint expression of the Persian poet Jami—"that the noble disposition of readers will induce them to pass over defects."[3]

(1) Of course, after I had assured them that, in my opinion, the portions to be used were entirely free from matter to which exception could be taken.

(2) In the spelling of Arabic words I have, as this is a Life of Burton, followed Burton, except, of course, when quoting Payne and others. Burton always writes 'Abu Nowas,' Payne 'Abu Nuwas,' and so on.

(3) Conclusion of *The Beharistan*.

My grateful thanks are due to the following ladies and gentlemen for various services.

Arbuthnot, Mrs. F. F., 43, South Street, Park Lane, London.
Ashbee, Mr. C. G., Woolstapler Hall, Chipping Cambden, Gloucestershire.
Agg, Mrs. Hewletts, Cheltenham.
Baddeley, Mrs., Brighton.
Baker, Dr. Grenfell, 25, Southwick Street, Hyde Park, W
Birch, Mrs. G. M., Lympstone Grange, South Devon.
Blumhardt, Prof. James F., British Museum.
Burton, Mrs. E. J., 31 Wilbury Road, Brighton.
Burton, Major St. George, The Black Watch.
Burton, Mr. Frederick, Brighton.
Cautley, Mr. P. P., 4, Via della Zonta, Trieste.
Clayton, Mr. Arthur South View Ropley Hants.
Carrington, Mr. Charles, 13, Faubourg Montmartre, Paris.
Chatto, Mr. Andrew, Hillside Elstree.
Codrington, Dr., Royal Asiatic Society Albemarle Street.
Committee, The, of the Central Library. Camberwell.
Eales, Rev. A. R. T., The Rectory, Elstree Herts.
Ellis, Mr. A. G., British Museum.
Editors, The, of the following newspapers : *The Times, The Daily Telegraph, The Standard, The Daily News, The Morning Post, The Daily Chronicle The Daily Mail, The Athenæum, The Saturday Review, The Academy*, for inserting letters for me at different times. These letters put me in touch with several of Burton's old friends.
Gardiner, Mr. C. H., 4, Montpelier Crescent, Brighton.
George, Mr. William H., 2, Highfield Terrace, Bognor.
Hector, Mr. E., Bookseller, 103, John Bright Street, Birmingham.
Hutchinson & Co., Messrs., for the loan of the portrait of Khamoor
Jones, Mr. Herbert, The Library, High Street Kensington.
Kirby, Mr. W. F., "Hilden," Sutton Court Road, Chiswick, London.
Letchford, Miss Daisy (now Madame Nicastro) Mezellina 178, Naples.
McCarthy, The Very Rev P. J. Canon, Ilkeston, Derbyshire.
Mendelssohn, Mr. S., 21, Kensington Court Gardens, London, W
Murray, Mr. T. Douglas, Pyt Cottage, Tisbury Wilts.
MacRitchie, Mr. David, 4, Archibald Place, Edinburgh.
Newcombe, Mr. C. F., 16, Champion Park, Denmark Hill, London, S.E.
Nicastro, Madame.
Payne, Mr. John.
Pelham, Dr., President of Trinity College, Oxford.
Pryce, Mr. E. S. Mostyn, Gunley Hall, Chirbury, Shropshire.
Rankin-Lloyd, Mrs., Wilne House, Pembroke.
Royal Asiatic Society (for permission to examine the Arbuthnot and Rehatsek manuscripts).
Roe, Rev. Henry, 12, Barnoon Terrace, St. Ives, Cornwall.
Sams, Rev. G. F., The Rectory, Emberton, Bucks.
Snowsill, Mr. W. G., Camberwell Central Library
Spencer, Mr. W. T., Bookseller, 27 New Oxford Street, London, W.C.
Segrave, Mr. H., Seaview, Lyme Regis, Dorset.
Tussaud, Mr. John, of "Madame Tussaud's."

Tedder, Mr., The Athenæum.
Tuckey, Dr. Charles Lloyd, 88, Park Street, Grosvenor Square, London.
Van Zeller, Mrs. (Lady Burton's sister).
Wilkins, Mr. W. H., 3, Queen Street, Mayfair, London, W.
Wood, Mr. W. Martin, Underwood, Oatlands Avenue, Weybridge.
Wyllie, Mr. Francis R. S., 6, Montpellier Villas, Brighton.
My wife, too, upon whom devolved the heavy task of transcribing, must also be awarded her meed of praise.

The following is a fairly complete list of the various Books and Magazine Articles that have been laid under contribution.

Arbuthnot, F. F., "Persian Portraits." 1887
"The Mysteries of Chronology."
"Life of Balzac (in Manuscript).
"Baily's Monthly Magazine," April 1883.
Baddeley, St. Clair (See Richards, A.B.)
Burton, Lady, "Life of Sir Richard Burton," 2 vols. 1893.
Her Works. 5 vols.
Burton, Sir Richard. His Works. 60 vols.
"Edinburgh Review," July 1886. No. 335.
Hitchman, F., "Richard R. Burton," 2 vols. 1887.
Kama Shastra Society's Publications.
Magazine Articles by or relating to Burton. Too numerous to mention.
Payne, Mr. John, The Book of "The Thousand Nights and One Night," 9 vols., 1882-4, and "Omar Kheyyam."
"Perfumed Garden, The." Published in 1904 by Mr. Carrington, of Paris.
Its Preface contains letters from several of the leading Arabists of the day, including M. Fagnan and Professor Hartwig Derenbourg, *Membre de l'Institut.*
Richards (A. B.), Wilson (A.), and Baddeley (St. C.), "Sketch of the Career of Richard F. Burton," 1886.
Rehatsek (Edward), Translations.
Roe, Rev. Henry, "West African Scenes," "Fernando Po Mission."
Stisted, Miss Georgiana, "Reminiscences of Sir Richard Burton"—"Temple Bar," July, 1891. Vol. 92.
"The True Life of Sir Richard Burton," 1896.
"Saturday Review,""Ultima Thule," 1876, Jan. 15 (p. 82).
"Zanzibar," 1872, "February 17th (p. 222).
Wilkins, W. H., "The Romance of Isabel Lady Burton," 2 vols. 1897.
Also the various works by Sir Richard Burton that have been edited by Mr. Wilkins.
Wilson, A. (See Richards, A. B.)

THOMAS WRIGHT.

CONTENTS OF VOLUME I

CHAPTER I

19th March 1821—October 1840

CHILDHOOD AND YOUTH

CHAPTER II

October 1840—April 1842

OXFORD

CHAPTER III.

April 1842—20th February 1847

SIND.

CHAPTER IV

20th February 1847—1849

UNDER THE SPELL OF CAMOENS

CHAPTER V

1849—3rd April 1853

CHIEFLY BOULOGNE

CHAPTER VI

3rd April 1853—29th October 1854

PILGRIMAGE TO MECCA

CHAPTER VII

29th October 1854—22nd April 1855

TO HARAR

CHAPTER VIII

22nd April 1855—December 1856

THE CRIMEA

CHAPTER IX

December 1856—21st May 1859

THE UNVEILING OF ISIS

CHAPTER X

21st May 1859—August 1861

MORMONS AND MARRIAGE

CHAPTER XI

August 1861—29th November 1863

FERNANDO PO

CHAPTER XII

29th November, 1863—10th November 1865

GELELE

CHAPTER XIII

October 1865—October 1869

2ND CONSULATE: SANTOS

CHAPTER XIV

1st October 1869—16th August 1871

"EMPEROR AND EMPRESS OF DAMASCUS"

CHAPTER XV

16th August 1871—4th June 1872

"THE BLACKNESS OF DARKNESS"

CHAPTER XVI

4th June 1872—24th October 1872

IN ICELAND

CHAPTER XVII

24th October 1872—12th May 1875

4TH CONSULATE : TRIESTE

CHAPTER XVIII

12th May 1875—18th June 1876

THE TRIP TO INDIA

CHAPTER XIX

18th June 1876—31st March 1877

COLONEL GORDON.

CHAPTER XX

31st March 1877—27th December 1879

MIDIAN

LIST OF PLATES IN VOLUME I

THE LIFE OF SIR RICHARD BURTON

CHAPTER I

19TH MARCH 1821—OCTOBER 1840

CHILDHOOD AND YOUTH

1. Torquay and Elstree.

Sir Richard Burton, the famous traveller, linguist, and anthropologist—"the Arabian Knight"—"the last of the demi-gods"—has been very generally regarded as the most picturesque figure of his time, and one of the most heroic and illustrious men that "this blessed plot . . . this England," this mother of heroes ever produced.

The Burtons, a Westmoreland family (1) who had settled in Ireland, included among their members several men of eminence, not only in the army, which had always powerfully attracted them, but also in the navy and the church. (2) For long there was a baronetcy in the family, but it fell into abeyance about 1712, and all attempts of the later Burtons to substantiate their claim to it proved ineffectual. (3)

Burton supposed himself to be descended from Louis XIV. La Belle Montmorency, a beauty of the French court, had, it seems, a son, of which she rather believed

(1) They came from Shap.

(2) Thus there was a Bishop Burton of Killala and an Admiral Ryder Burton. See Genealogical Tree in the Appendix.

(3) Mrs. Burton made a brave attempt in 1875, but could never fill the gap between 1712 and 1750.

Louis to be the father. In any circumstances she called the baby Louis Le Jeune, put him in a basket of flowers and carried him to Ireland, where he became known as Louis Drelincourt Young. Louis Young's grand-daughter married the Rev. Edward Burton, Richard Burton's grandfather. Thus it is possible that a runnel of the blood of "le grand monarque" tripped through Burton's veins. But Burton is a Romany name, and as Richard Burton had certain gipsy characteristics, some persons have credited him with gipsy lineage. Certainly no man could have been more given to wandering. Lastly, through his maternal grandmother, he was descended from the famous Scotch marauder, Rob Roy.

Burton's parents were Lieutenant-Colonel Joseph Netterville Burton, a tall, handsome man with sallow skin, dark hair, and coal-black eyes, and Martha Beckwith, the accomplished but plain daughter of Richard and Sarah Baker, of Barham House (now "Hillside"[1]), Elstree, Hertfordshire.

Richard Baker was an opulent country gentleman, and the most important personage in the parish. Judging from the size of his pew at church, "No. 19," he must also have been a man of eminent piety, for it contained sixteen sittings. At all events he kept the parish in admirable order, and, as churchwarden, discountenanced unreasonable sleeping in church. Thanks to his patronage the choir made marked progress, and eventually there was no louder in the county. In 1813 we find him overseer with one George Olney. He took a perfunctory[2] interest in the village school (where, by the by, Arthur Orton, the Tichborne claimant, received his elaborate education), and was for a time "director." He led the breezy life of a country gentleman. With his fat acres,

(1) Now the residence of Mr. Andrew Chatto, the publisher.

(2) In 1818 the Inspector writes in the Visitors' Book: "The Bakers seldom there." Still, the Bakers gave occasional treats to the children, and Mrs. Baker once made a present of a new frock to each of the girls.

ELSTREE CHURCH, 1823.

ELSTREE CHURCH

his thumping balance at the bank, his cellar of crusted wine, and his horse that never refused a gate, this world seemed to him a nether paradise. He required, he said, only one more boon to make his happiness complete—namely, a grandson with unmistakably red hair. A shrewd man of business, Mr. Baker tied up every farthing of his daughter's fortune, £30,000; and this was well, for Burton's father, a rather Quixotic gentleman, had but a child's notion of the use of money. The Burtons resided at Torquay, and Colonel Burton busied himself chiefly in making chemical experiments, of which he was remarkably fond; but the other members of the household, who generally went about holding their noses, appear not to have sympathised with his studies and researches. He was very superstitious—nothing, for instance, could induce him to reveal his birthday; and he fretted continually because he was not permitted to invest his wife's money and make a second fortune; which no doubt he would very soon have done — for somebody else.

Richard Francis Burton was born at Torquay [1] on 19th March 1821; and to the intemperate joy of the family his hair was a fierce and fiery red. The news flew madly to Elstree. Old Mr. Baker could scarcely contain himself, and vowed then and there to leave the whole of his fortune to his considerate grandson. The baby, of course, was promptly called Richard after Mr. Baker, with Francis as an afterthought; and a little later the Burtons went to reside at Barham House with the grandparents. Richard was baptised in the parish church at Elstree, 2nd September 1821. In the entry his father's abode is called "Bareham Wood," [2] the name being spelt various ways. Our illustration of the old church is taken from an engraving made to commemorate the burial of William

(1) Not at Elstree as Sir Richard Burton himself supposed and said, and as all his biographers have reiterated. It is plainly stated in the Elstree register that he was born at Torquay.

(2) The clergyman was David Felix.

Weare [1] murdered by the notorious John Thurtell; an event that occurred in 1823, when Burton was two years old.

There was another link between the Burtons and the Bakers, for Joseph Netterville's youngest brother, Francis, military surgeon in the 66th regiment, married Sarah Baker, Mr. Richard Baker's eldest daughter. Dr. Burton [2] who was in St. Helena at the time of Napoleon's death lives in history as the man who "took a bust of the dead emperor." [3]

Being subject to asthma, Colonel Burton now left England and hired a château called Beauséjour situate on an eminence near Tours, where there was an English colony. For several years the family fluctuated between Tours and Elstree, and we hear of a great yellow chariot which from time to time rolled into daylight. Richard's hair gradually turned from its fiery and obtrusive red to jet black, but the violent temper of which the former colour is supposed to be indicative, and of which he had already many times given proofs, signalised him to the end of life. In 1823 Mrs. Burton gave birth to a daughter, Maria. Katharine Elisa, who became the wife of General Sir Henry Stisted; and on 3rd July 1824 to a son, Edward Joseph Netterville, both of whom were baptized at Elstree.[4] While at Tours the children were under the care of their Hertfordshire nurse, Mrs. Ling, a good, but obstinately English soul who had been induced to cross the Channel only after strenuous opposition.

2. Tours and Elstree.

Richard Burton always preserved some faint recollections of his grandfather. "The first thing I remember,"

(1) Weare's grave is unmemorialled, so the spot is known only in so far as the group in the picture indicates it.

(2) He died 24th October 1828, aged 41; his wife died 10th September 1848. Both are buried at Elstree church, where there is a tablet to their memory.

(3) For a time Antommarchi falsely bore the credit of it.

(4) Maria, 18th March 1823; Edward, 31st August 1824.

ELSTREE VILLAGE.

BARHAM HOUSE, ELSTREE, WHERE BURTON SPENT HIS CHILDHOOD.

MRS. BAKER.

FROM THE ORIGINAL IN THE POSSESSION OF MRS. AGG, OF CHELTENHAM.

Photo by J. Holloway, Cheltenham.

MR. RICHARD BAKER.

FROM THE ORIGINAL IN THE POSSESSION OF MRS. AGG, OF CHELTENHAM.

Photo by J. Holloway, Cheltenham

he says, "was being brought down after dinner at Barham House to eat white currants, seated upon the knee of a tall man with yellow hair and blue eyes." This would be in the summer of 1824. Mr. Baker, as we have seen, had intended to leave the whole of his property—worth about half a million—to his red-haired grandson; and an old will, made in 1812, was to be cancelled. But Burton's mother had a half brother—Richard Baker, junior—to whom she was extravagantly attached, and, in order that this brother should not lose a fortune, she did everything in her power to prevent Mr. Baker from carrying out his purpose Three years passed away, but at last Mr. Baker resolved to be thwarted no longer, so he drove to his lawyer's. It was the 16th of September 1824. He reached the door and leapt nimbly from his carriage; but his foot had scarcely touched the ground before he fell dead of heart disease. So the old will had to stand, and the property, instead of going to Burton, was divided among the children of Mr. Baker, Burton's mother taking merely her share. But for this extraordinary good hap Richard Burton might have led the life of an undistinguished country gentleman; ingloriously breaking his dogs, training his horses and attending to the breed of stock. The planting of a quincunx or the presentation of a pump to the parish might have proved his solitary title to fame. Mr. Baker was buried at Elstree church, where may be seen a tablet to him with the following inscription:

3. Death of Richard Baker, 16th September 1824.

> "Sacred to the memory of Richard Baker, Esq., late of Barham House in this parish, who departed this life on the 16th September 1824, aged 62 years." (1)

Soon after the death of her husband Mrs Baker must have left Elstree,(2) for from 1827 to 1839, Barham House

(1) Beneath is an inscription to his widow, Sarah Baker, who died 6th March, 1846, aged 74 years.

(2) Her last subscription to the school was in 1825. In 1840 she lived in Cumberland Place, London.

was occupied by Viscount Northland. The Burtons continued to reside at Tours, and all went well until cholera broke out. Old Mrs. Baker, hearing the news, and accounting prevention better than cure, at once hurried across the channel; nor did she breathe freely until she had plugged every nose at Beauséjour with the best Borneo camphor.

The apprehensive old lady, indeed, hovered round her grandchildren all day like some guardian angel, resolutely determined that no conceivable means should be spared to save them from the dreaded epidemic; and it was not until she had seen them safely tucked in their snowy, lavendered beds that her anxieties of the day really ceased. One night, however, when she went, as was her custom, to look at the sleeping children before retiring herself, she found, to her horror, that they were not there. The whole household was roused, and there was an agonising hue and cry; but, by and by, the culprits were seen slinking softly in at the principal door. It seems that they had climbed down from their room and had gone the round with the death carts and torches, to help collect corpses; and enquiry revealed that they had worked considerably harder than the paid men. When the cholera scare passed off Mrs. Baker took to learning French, and with such success than in less than six months she was able to speak several words, though she could never get hold of the correct pronunciation. Despite, however, her knowledge of the language, the good lady did not take kindly to France, and she often looked wistfully northwards, quoting as she did so her favourite Cowper

"England with all thy faults I love thee still."

She and Mrs. Ling, the old nurse, who pined for English beef and beer, made some attempts to console each other, but with inappreciable success, and finally the fellow-sufferers, their faces now beaming with smiles, returned together to their England. And not even Campbell's sailor lad was gladder to see again the "dear cliffs of Dover."

RICHARD BURTON, HIS SISTER MARIA AND BROTHER EDWARD

FROM A PAINTING IN THE POSSESSION OF MRS. AGG, OF CHELTENHAM.

Photo by J. Holloway, Cheltenham.

Our charmingly quaint picture of Richard, his sister and brother, in wondrous French costumes, is from an oil painting [1] which has not before been copied. Richard was first taught by a lame Irishman named Clough, who kept a school at Tours; and by and by, chiefly for the children's sake, Colonel Burton gave up Beauséjour and took a house in the Rue de l'Archevêché, the best street in the town. The little Burtons next attended the academy of a Mr. John Gilchrist, who grounded them in Latin and Greek. A kind-hearted man, Mr. Gilchrist often gave his pupils little treats. Once, for instance, he took them to see a woman guillotined. Richard and Edward were, to use Richard's expression, "perfect devilets." Nor was the sister an angelet. The boys lied, fought, beat their maids, generally after running at their petticoats and upsetting them, smashed windows, stole apple puffs; and their escapades and Richard's ungovernable temper were the talk of the neighbourhood. Their father was at this time given to boar hunting in the neighbouring forest, but as he generally damaged himself against the trees and returned home on a stretcher, he ultimately abandoned himself again to the equally useful but less perilous pursuit of chemistry. If Colonel Burton's blowpipes and retorts and his conduct in private usually kept Mrs. Burton on tenterhooks, she was no less uneasy on his account when they went into society. He was so apt to call things by their right names. Thus on one occasion when the conversation ran upon a certain lady who was known to be unfaithful to her husband, he inexpressibly shocked a sensitive company by referring to her as "an adulteress." In this trait, as in many others, his famous son closely resembled him.

A youthful Stoic, Burton, in times of suffering, invariably took infinite pains to conceal his feelings. Thus all one day he was in frightful agony with the toothache, but nobody else knew anything about it until next morning, when his cheek was swollen to the size of a peewit's egg.

(1) The original is now in the possession of Mrs. Agg, of Cheltenham.

He tried, too, to smother every affectionate instinct; but when under strong emotion was not always successful. One day, throwing stones, he cut his sister's forehead. Forgetting all his noble resolutions he flew to her, flung his arms round her, kissed her again and again, and then burst into a fit of crying. Mrs. Burton's way of dressing her children had the charm of simplicity. She used to buy a piece of yellow nankin and make up three suits as nearly as possible alike, except for size. We looked, said Burton, "like three sticks of barley sugar," and the little French boys who called after them in the streets thought so too, until Richard had well punched all their heads, when their opinions underwent a sudden change.

Another household incident that fixed itself in Burton's mind was the loss of their "elegant and chivalrous French chef," who had rebelled when ordered to boil a *gigot*. "*Comment, madame*," he replied to Mrs. Burton, "*un—gigot!—cuit à l'eau, jamais! Neverre!*" And rather than spoil, as he conceived it, a good leg of mutton he quitted her service.[1] Like most boys, Burton was fond of pets, and often spent hours trying to revive some bird or small beast that had met with misfortune, a bias that affords a curious illustration of the permanence of character. The boy of nine once succeeded in resuscitating a favourite bullfinch which had nearly drowned itself in a great water jug—and we shall find the man of sixty-nine, on the very last day of his life, trying to revive a half-drowned robin.

In 1829 the Burtons returned to England and took a house in Maids of Honour Row, Richmond, while Richard and Edward were sent to a preparatory school at Richmond Green—a handsome building with a paddock which enclosed some fine old elms—kept by a "burly savage," named the Rev. Charles Delafosse. Although the fees were high, the school was badly conducted, and the boys were both ill-taught and ill-fed. Richard employed himself out of

4. At School, Richmond, 1829.

[1] *Wanderings in West Africa*, ii. p. 143.

school hours fighting with the other boys, and had at one time thirty-two affairs of honour to settle. "On the first occasion," he says, "I received a blow in the eye, which I thought most unfair, and having got my opponent down I proceeded to hammer his head against the ground, using his ears by way of handles. My indignation knew no bounds when I was pulled off by the bystanders, and told to let my enemy stand up again. 'Stand up!' I cried, 'After all the trouble I've had to get the fellow down.'"(1)

Of the various countries he knew, Burton hated England most. Would he ever, he asked, see again his "Dear France." And then Fate, who revels in irony, must needs set him to learn as a school task, of all the poems in English, Goldsmith's *Traveller!* So the wretched boy, cursing England in his heart, scowling and taking it out of Goldsmith by daubing his pages with ink, sat mumbling:

> "Such is the patriot's boast, where'er we roam,
> His first, best country ever is at home." (2)

By and by, to Burton's extravagant joy—and he always intemperately loved change—measles broke out in the school, the pupils were dispersed, and Colonel Burton, tired of Richmond, resolved to make again for the continent. As tutor for his boys he hired an ox-like man "with a head the shape of a pear, smaller end uppermost"—the Rev. H. R. Du Pré afterwards rector of Shellingford; and Maria was put in charge of a peony-faced lady named Miss Ruxton. The boys hurrahed vociferously when they left what they called wretched little England; but subsequently Richard held that his having been educated abroad was an incalculable loss to him. He said the more English boys are, "even to the cut of their hair," the better their chances in life. Moreover, that it is a real advantage to belong to some parish. "It is a great thing when you have won a battle, or explored

(1) *Life* i. 29.

(2) Goldsmith's *Traveller*, lines 73 and 74.

Central Africa, to be welcomed home by some little corner of the great world, which takes a pride in your exploits, because they reflect honour on itself.''[1] An English education might have brought Burton more wealth, but for the wild and adventurous life before him no possible training could have been better than the varied and desultory one he had. Nor could there have been a more suitable preparation for the great linguist and anthropologist. From babyhood he mixed with men of many nations.

5. The Continent Again.

At first the family settled at Blois, where Colonel and Mrs. Burton gave themselves over to the excitement of dressing three or four times a day; and, as there was nothing whatever the matter with them, passed many hours in feeling each other's pulses, looking at each other's tongues, and doctoring each other. Richard and Edward devoted themselves to fencing and swimming. If the three children were wild in England they were doubly wild at Blois. Pear-headed Mr. Du Pré stuck tenaciously to his work, but Miss Ruxton gave up in despair and returned to England. At a dancing party the boys learnt what it was to fall in love. Richard adored an extremely tall young woman named Miss Donovan, "whose face was truly celestial—being so far up" but she was unkind, and did not encourage him.

After a year at Blois, Colonel and Mrs. Burton, who had at last succeeded in persuading themselves that they were really invalids, resolved to go in search of a more genial climate. Out came the cumbersome old yellow chariot again, and in this and a chaise drawn by an ugly beast called Dobbin, the family, with Colonel Burton's blowpipes, retorts and other "notions," as his son put it, proceeded by easy stages to Marseilles, whence chariot, chaise, horse and family were shipped to Leghorn, and a few days later they found themselves at Pisa. The boys became proficient in Italian and drawing, but it was not until

(1) *Life*, i. 32.

middle life that Richard's writing developed into that gossamer hand which so long distinguished it. Both had a talent for music, but when "a thing like Paganini, length without breadth" was introduced, and they were ordered to learn the violin, Richard rebelled, flew into a towering rage and broke his instrument on his master's head. Edward, however, threw his whole soul into the work and became one of the finest amateur violinists of his day. Edward, indeed, was the Greek of the family, standing for music and song as well as for muscle. He had the finely chiselled profile and the straight nose that characterises the faces on Attic coins. Richard, though without the Roman features, was more of the ancient Roman type of character: severe, doggedly brave, utilitarian; and he was of considerably larger mould than his brother. In July 1832, the family stayed at Siena and later at Perugia, where they visited the tomb of Pietro Aretino. At Florence, the boys, having induced their sister to lend them her pocket money, laid it out in a case of pistols; while their mother went in daily terror lest they should kill each other. The worst they did, however, was to put a bullet through a very good hat which belonged to Mr. Du Pré. When their mother begged them not to read Lord Chesterfield's *Letters to a Son*, concerning the morality of which she had doubts, they dutifully complied and surrendered themselves piously, and without a murmur, to the chaste pages of Paul de Kock. They did not, however, neglect the art treasures of Florence; and at Rome, their next stopping-place, they sauntered about with Baedeker's predecessor, "Mrs. Starke," and peered into earthy churches and flower-illumined ruins. Later the family journeyed to Naples, where the boys continued their studies under Mr. Du Pré. As a clergyman, this gentleman steadily inculcated in his pupils the beautiful principles of the Christian religion, and took a sincere and lively interest in their favourite pastime of cock-fighting.

Colonel Burton continued his chemical studies, and

in an evil hour for the family, purchased a copy of the quaint text book by S. Parkes: "A Chemical Catechism . . . . with copious notes . . . . to which are added a Vocabulary and a Chapter of Amusing Experiments."[1] And very amusing they were when Colonel Burton made them. Having studied the book closely, including the "poetry" with which it is studded, he manufactured, at vast expense, a few cakes of a nasty-looking and evil-smelling substance which, he said, was soap, and ought to be put on the market. Mrs. Burton intimated that he might put it on the market or anywhere else as long as he did not make any more. He next, by the aid of the same manual, prepared a mixture which he called citric acid, though any other name would have suited it equally well; and of this, as neither he nor anybody else had any use for it, he daily produced large quantities. From Naples the family moved to Sorrento, where S'or Riccardo and S'or Edwardo, as the Italians called them, surrendered themselves to the natural and legendary influences of the neighbourhood and to reading. The promontory on which Sorrento stands is barren enough, but southward rise pleasant cliffs viridescent with samphire, and beyond them purple hills dotted with white spots of houses. At no great distance, though hidden from view, stood the classic Paestum, with its temple to Neptune; and nothing was easier than to imagine, on his native sea as it were, the shell-borne ocean-god and old Triton blowing his wreathèd horn. Capri, the retreat of Tiberius, was of easy access. Eastward swept a land of myrtle and lemon orchards. While the elder Burton was immersed in the melodious Parkes, who sang about "Oxygen, abandoning the mass," and changing "into gas," his sons played the parts of Anacreon and Ovid, they crowned their heads with garlands and drank wine like Anacreon, not omitting the libation, and called to mind the Ovid of well-nigh two thousand years previous,

(1) It seems to have been first issued in 1807. There is a review of it in *The Anti-Jacobin* for that year.

and his roses of Paestum. From poetry they turned once more to pistols, again brought their mother's heart to her mouth, and became generally ungovernable. A visit to a house of poor reputation having been discovered, their father and Mr. Du Pré set upon them with horsewhips, whereupon the graceless but agile youths ran to a neighbouring house and swarmed to the top of a stack of chimneys, whence partly by word and partly by gesticulation they arranged terms of peace.

In 1836, the Burtons left for Pau in the South of France; and while there Richard lost his heart to the daughter of a French baron. Unfortunately, however, she had to go away to be married; and Richard who loved her to desperation, wept bitterly, partly because he was to lose her and partly because she didn't weep too. Edward and the young lady's sister, who also understood each other, fared no better, for Colonel Burton having got tired of Pau, the whole family had to return to Italy. At Pisa "S'or Riccardo" and S'or Edwardo" again "cocked their hats and loved the ladies," Riccardo's choice being a slim, soft, dark beauty named Caterina, Edwardo's her sister Antonia. Proposals of marriage were made and accepted, but adieux had soon to follow, for Colonel Burton now moved to Lucca. All four lovers gave way to tears, and Richard was so wrung with grief that he did not become engaged again for over a fortnight. At Lucca the precious pair ruffled it with a number of dissolute medical students, who taught them several quite original wickednesses. They went, however, with their parents, into more wholesome society; and were introduced to Louis Desanges, the battle painter, Miss Helen Croly, daughter of the author of *Salathiel*, and Miss Virginia Gabriel (daughter of General, generally called Archangel Gabriel) the lady who afterwards attained fame as a musical composer[1] and became, as we have recently discovered, one of the friends of Walter Pater. Says

(1) She was thrown from her carriage, 7th August 1877, and died in St. George's Hospital.

Burton " she showed her *savoir faire* at the earliest age. At a ball given to the Prince, all appeared in their finest dresses, and richest jewellery. Miss Virginia was in white, with a single necklace of pink coral." They danced till daybreak, when Miss Virginia "was like a rose among faded dahlias and sunflowers."

Here, as everywhere, there was more pistol practice, and the boys plumed themselves on having discovered a new vice—that of opium-eating, while their father made the house unendurable by the preparation of sulphuretted hydrogen and other highly-scented compounds. It was recognised, however, that these chemical experiments had at least the advantage of keeping Colonel Burton employed, and consequently of allowing everybody a little breathing time at each stopping place. In the spring of 1840, Colonel Burton, Mr. Du Pré and the lads set out for Schinznach, in Switzerland, to drink the waters; and then the family returned to England in order that Richard and Edward might have a university education. Their father, although not quite certain as to their future, thought they were most adapted for holy orders. Their deportment was perfect, the ladies admired them, and their worst enemies, it seems, had never accused them of being "unorthodox in their views." Indeed, Mrs. Burton already pictured them mitred and croziered. For a few weeks the budding bishops stayed with "Grandmama Baker," who with "Aunt Sarah" and "Aunt Georgiana," and Aunt Sarah's daughters, Sarah and Elisa, was summering at Hampstead; and filled up the time, which hung heavy on their hands, with gambling, drinking, and love-making

TRINITY COLLEGE, OXFORD.

From a Photograph by Hills and Saunders, Oxford.

CHAPTER II

OCTOBER 1840—APRIL 1842

OXFORD

Edward was then placed under a clergyman at Cambridge—The Rev. Mr. Havergal, whose name, to that gentleman's indignation, the brothers turned into "a peculiar form of ridicule." (1) Richard was to go to Trinity College, Oxford. Neither, as we have seen, had been suitably prepared for a University career. Richard, who could speak fluently French, Italian, and modern Greek, did not know the Apostles' Creed, and what was even more unusual in a prospective clergyman, had never heard of the Thirty-nine Articles. He was struck with the architecture of the colleges, and much surprised at the meanness of the houses that surrounded them. He heretically calls the Isis 'a mere moat,' the Cherwell 'a ditch.' The brilliant dare-devil from Italy despised alike the raw limitary, reputable, priggish undergraduates and the dull, snuffling, smug-looking, fussy dons. The torpor of academic dulness, indeed, was as irksome to Burton at Oxford as it had been to FitzGerald and Tennyson at Cambridge. After a little coaching from Dr. Ogle and Dr. William Alexander Greenhill, (2) he in October 1840, entered Trinity, where he was installed in "a couple of frowsy dog-holes" overlooking the garden of old Dr. Jenkins, the Master of Balliol.

6. Trinity College, October 1840.

"My reception at College," says Burton, "was not

(1) *Life*, by Lady Burton, i. 67.

(2) Dr. Greenhill (1814—1894), physician and author of many books.

pleasant. I had grown a splendid moustache, which was the envy of all the boys abroad, and which all the advice of Drs. Ogle and Greenhill failed to make me remove. I declined to be shaved until formal orders were issued by the authorities of the college. For I had already formed strong ideas upon the Shaven Age of England, when her history, with some brilliant exceptions, such as Marlborough, Wellington and Nelson, was at its meanest." An undergraduate who laughed at him he challenged to fight a duel; and when he was reminded that Oxford men" liked to visit freshmen's rooms and play practical jokes, he stirred his fire, heated his poker red hot, and waited impatiently for callers. "The college teaching for which one was obliged to pay," says Burton, "was of the most worthless description. Two hours a day were regularly wasted, and those who read for honours were obliged to choose and pay a private coach."

Another grievance was the constant bell ringing, there being so many churches and so many services both on week days and Sundays. Later, however, he discovered that it is possible to study, even at Oxford, if you plug your ears with cotton-wool soaked in glycerine. He spent his first months, not in studying, but in rowing, fencing, shooting the college rooks, and breaking the rules generally. Many of his pranks were at the expense of Dr. Jenkins, for whose sturdy common sense, however, he had sincere respect; and long after, in his *Vikram and the Vampire*, in which he satirises the tutors and gerund-grinders of Oxford, he paid him a compliment. (1)

Although he could not speak highly of the dons and undergraduates, he was forced to admit that in one respect the University out-distanced all other seats of learning. It produced a breed of bull-terriers of renowned pedigree which for their "beautiful build" were a joy to think about and a delirium to contemplate; and of one of these pugnacious brutes he soon became

(1) *Vikram and the Vampire*, Seventh Story, about the pedants who resurrected the tiger.

the proud possessor. That he got drunk himself and made his fellow collegians drunk he mentions quite casually, just as he mentions his other preparations for holy orders. If he walked out with his bull-terrier, it was generally to Bagley Wood, where a pretty, dizened gipsy girl named Selina told fortunes; and henceforward he took a keen interest in Selina's race.

He spent most of his time, however, in the fencing saloons of an Italian named Angelo and a Scotchman named Maclaren; and it was at Maclaren's he first met Alfred Bates Richards, who became a life friend. Richards, an undergraduate of Exeter, was a man of splendid physique. A giant in height and strength, he defeated all antagonists at boxing, but Burton mastered him with the foil and the broad-sword. Richards, who, like Burton, became a voluminous author, [1] wrote long after, "I am sure, though Burton was brilliant, rather wild, and very popular, none of us foresaw his future greatness."

Another Oxford friend of Burton's was Tom Hughes, author of *Tom Brown's Schooldays*; the man who, in Burton's phrase, "taught boys not to be ashamed of being called good,"[2] and he always revered the memory of his tutor, the Rev. Thomas Short. [3] Burton naturally made enemies as well as friends, but the most bitter was that imaginary person, Mrs. Grundy. This lady, whom he always pictured as an exceedingly stout and square-looking body with capacious skirts, and a look of austere piety, had, he tells us, "just begun to reign" when he was at Oxford, although forty years had elapsed since she first made her bow, [4] and set everybody asking, "What will Mrs. Grundy say?" Mrs. Grundy had a great deal to say against Richard Burton, and, life through, he took a

(1) He edited successively *The Daily Telegraph* and *The Morning Advertiser*, wrote plays and published several volumes of poetry. He began *The Career of R. F. Burton*, and got as far as 1876.

(2) *City of the Saints*, p. 513.

(3) Short died 31st May 1879, aged 90.

(4) In Thomas Morton's Play *Speed the Plough*, first acted in 1800

peculiar delight in affronting her. The good soul disapproved of Burton's "foreign ways" and his "expressed dislike to school and college life," she disapproved of much that he did in his prime, and when he came to translate *The Arabian Nights* she set up, and not without justification, a scream that is heard even to this day and in the remotest corners of the kingdom.

If Richard was miserable at Oxford, Edward was equally so at Cambridge. After the polish and politeness of Italy, where they had been "such tremendous dandies and ladies' men," the boorishness and shoppiness" of Oxford and Cambridge were well-nigh unendurable. Seizing an early opportunity, Richard ran over to Cambridge to visit his brother. "What is the matter, Edward," enquired Richard. "Why so downcast?" "Oh, Dick, "moaned Edward, "I have fallen among *épiciers*.[1]"

7. Expelled, April 1842.

The dull life at Oxford was varied by the occasional visit of a mesmeric lecturer; and one youth caused peals of canorous laughter by walking round in a pretended mesmeric sleep and kissing the pretty daughters of the dons.

The only preacher Burton would listen to was Newman, then Vicar of St. Mary's; of Pusey's interminable and prosy harangues he could not bear even to think. Although unable to bend himself to the drudgery of Oxford, Burton was already forming vast ambitions. He longed to excel as a linguist, and particularly in Oriental languages. Hence he began to teach himself Arabic; and got a little assistance from the Spanish scholar Don Pascual de Gayangos. When he asked the Regius Professor of Arabic to teach him, he was rebuffed with the information that it was the duty of a professor to teach a class, not an individual. He spent the vacation with his Grandmother Baker in Great Cumberland Place, and he and his brother amused themselves about town with other roisterers, chiefly in gambling. Returned to

(1) Grocers.

Oxford he applied himself sedulously to the acquisition of foreign languages. He says, "I got a simple grammar and vocabulary, marked out the forms and words which I knew were absolutely necessary, and learnt them by heart. . . I never worked more than a quarter of an hour at a time, for after that the brain lost its freshness. After learning some three hundred words, easily done in a week, I stumbled through some easy book-work and underlined every word that I wished to recollect. . . Having finished my volume, I then carefully worked up the grammar minutiæ, and I then chose some other book whose subject most interested me. The neck of the language was now broken, and progress was rapid. If I came across a new sound, like the Arabic *Ghayn*, I trained my tongue to it by repeating it so many thousand times a day. When I read, I invariably read out loud, so that the ear might aid memory. I was delighted with the most difficult characters, Chinese and Cuneiform, because I felt that they impressed themselves more strongly upon the eye than the eternal Roman letters."(1) Such remarks from the man who became the the first linguist of his day are well worth remembering. For pronouncing Latin words the "Roman way" he was ridiculed, but he lived long enough to see this pronunciation adopted in all our schools. The long vacation of 1841 was spent at Wiesbaden with his father and mother. Here again the chief delights of Richard and his brother were gambling and fencing; and when tired of Wiesbaden they wandered about the country, visiting among other places Heidelberg and Mannheim. Once more Richard importuned his father to let him leave Oxford and enter the army, but Colonel Burton, who still considered his son peculiarly fitted for the church, was not to be moved. Upon his return to England, however, Burton resolved to take the matter into his own hands. He laid his plans, and presently—in April 1842—an opportunity offered.

The Oxford races of that year were being looked

(1) *Life*, i. 81.

forward to with exceptional interest because of the anticipated presence of a noted steeplechaser named Oliver, but at the last moment the college authorities forbade the undergraduates to attend them.

Burton, however, and some other lawless spirits resolved to go all the same, and a tandem conveyed them from the rear of Worcester College to the race meeting. Next morning the culprits were brought before the college dignitaries ; but the dons having lectured Burton, he began lecturing them—concluding with the observation that young men ought not to be treated like children. As a consequence, while the other offenders were merely rusticated, Burton was expelled.[1] He made a ceremonious bow, and retired "stung with a sense of injustice," though where the injustice comes in, it is difficult to see. His departure from Oxford was characteristic. He and Anderson of Oriel, one of the other offenders, hired a tandem in which they placed their luggage, and then with "a cantering leader and a high-trotting shaft horse" they rode through the High Street, and so on to London, Burton artistically performing upon a yard of tin trumpet, waving adieux to his friends and kissing his hands to the shop girls. About the same time Edward, also for insubordination, had to leave Cambridge. Thus Burton got his own way, but he long afterwards told his sister, Lady Stisted, that beneath all his bravado there lay a deep sense of regret that such a course had been necessary.

(1) Or so he said. The President of Trinity writes to me: "He was repaid his caution money in April 1842. The probability is that he was rusticated for a period." If so, he could have returned to Oxford after the loss of a term or two.

CHAPTER III

APRIL 1842—20TH FEBRUARY 1847

SIND

On his arrival in London, Burton, in order to have an hour or two of peace, coolly told his people that he had been given an extra vacation, " as a reward for winning a double first." Then occurred a quite un-looked-for sequel. His father insisted on giving a dinner in honour of the success, and Burton, unwillingly enough, became the hero of the moment. At table, however, a remark from one of the guests revealed the precise truth—with the result of an unpleasant scene; but eventually it was deemed advisable to let Burton have his own way and exchange the surplice for the sword. The Indian Service having been selected, a commission was purchased for £500, and Burton presently found himself Ensign to the 18th Regiment, Bombay Native Infantry. Delirious with joy, he applied himself vigorously to Hindustani under a dirty, smoky Scotch linguist, named Duncan Forbes. While thus employed he made the acquaintance of two persons who just then enjoyed a remarkable reputation, namely John Varley,[1] the water colour painter and occultist, and the Rev. Robert Montgomery.[2] An artist of undoubted genius, Varley usually got fair prices for his pictures, but the expenses of a numerous family kept him miserably poor. Then he took to " judicial astrology," and eventually made it a kind of second

8. To Bombay, 18th June 1842.

(1) He died 17th November 1842, aged 65.

(2) Robert Montgomery 1807—1855.

profession. Curious to say, some of his predictions came true, and thanks to this freak of fate he obtained more fame from his horoscopes than from his canvasses. He "prognosticated," says Burton, "that I was to become a great astrologer." Straightway Burton buried himself in astrological and cabalistic books, [1] studied the uncanny arts, and became learned in "dark spells and devilish enginery," but his own prophecies generally proved to be of the Moseilima type; that is to say, the opposite invariably happened—a fatality that pursued him to the end of life. The Rev. Robert Montgomery, with whom also he became acquainted, was the fashionable preacher and author whom Macaulay cudgelled so pitilessly in the *Edinburgh Review*. Burton's aunts, Sarah and Georgiana, [2] who went with the crowd to his chapel, ranked the author of "Satan, a Poem," rather above Shakespeare, and probably few men have received higher encomiums or a greater number of wool-work slippers.

Having been sworn in at the East India House, Burton went down to Greenwich, whence on 18th June, 1842, after being "duly wept over," he, in company with his beautifully built bull-terrier of renowned pedigree, set sail for Bombay. He divided his time during the voyage, which lasted four months, between studying Hindustani and taking part in the quarrels of the crew. This was the year of the murder of Sir William Macnaughten by the Afghans and the disastrous retreat of the British from Cabul; consequently the first request of the voyagers on reaching Bombay (28th October 1842) was for news about Afghanistan. They learnt that the prestige of the British arms had been restored by Pollock, and that the campaign was ended.

To Burton, who had counted on being sent to the

(1) "My reading also ran into bad courses—Erpenius, Zadkiel Falconry, Cornelius Agrippa"—*Burton's Autobiographical Fragment*

(2) Sarah Baker (Mrs. Francis Burton), Georgiana Baker (Mrs. Bagshaw).

front, this was a burning disappointment. He found Bombay marvellously picturesque, with its crowds of people from all parts of the world, but before many days had passed he fell ill and had to be transferred to the Sanitarium, where he made the acquaintance of an old Parsee priest who assisted him in his Hindustani. Even in these early days we find him collecting material of the kind that was to be utilised in his *Arabian Nights*. He was struck, for example, with the fine hedges of henna whose powerful and distinctive odour loaded the atmosphere; and with the immense numbers of ravenous kites and grey-headed crows that swooped down on dead and even dying animals.

After six weeks' rest, having received orders to join his regiment, which was then stationed at Baroda, he engaged some Goanese servants and made the voyage thither in a small vessel called a pattymar. It took them four days to march from the Tankaria-Bunder mudbank, where they landed, to Baroda; and Burton thus graphically describes the scenery through which they passed. "The ground, rich black earth . . . . was covered with vivid, leek-like, verdigris green. The little villages, with their leafy huts, were surrounded and protected by hedge milk bush, the colour of emeralds. A light veil, as of Damascene silver, hung over each settlement, and the magnificent trees were tipped by peacocks screaming their good-night to the sun." The sharp bark of the monkey mingled with the bray of the conch. Arrived at Baroda, he lodged himself in a bungalow, and spent his time alternately there with his books and on the drill ground. He threw himself into his studies with an ardour scarcely credible—devoting twelve hours a day to Hindustani, and outwearying two munshis.

9. Baroda. The Bubu.

At that time it was quite the custom for the officers, married as well as single, to form irregular unions with the Hindu women. Every individual had his Bubu; consequently half-caste children were not uncommon;

but Burton was of opinion that this manner of life had advantages as well as disadvantages. It connected, he says, "the white stranger with the country and its people, gave him an interest in their manners and customs, and taught him thoroughly well their language." Like the rest, Burton had his Bubu. Still, he was no voluptuary. Towering ambition, enthusiasm, and passion for hard work trampled down all meaner instincts. Languages, not amours, were his aspiration, and his mind ran on grammar books rather than ghazels; though he confesses to having given whole days and nights to the tender pages of Euclid. Indeed, he was of a cold nature, and Plutarch's remark about Alexander applies equally to him: "For though otherwise he was very hot and hasty, yet was he hardly moved with lust or pleasure of the body." When the officers were not on the drill ground or philandering with their dusky loves, they amused themselves shooting the black buck, tigers, and the countless birds with which the neighbourhood abounded. The dances of the apish-looking Nautch girls, dressed though they were in magnificent brocades, gave Burton disgust rather than pleasure. The Gaikwar, whose state processions were gorgeous to a wonder, occasionally inaugurated spectacles like those of the old Roman arena, and we hear of fights between various wild animals. "Cocking" was universal, and Burton, who as a lad had patronised this cruel sport, himself kept a fighter—"Bhujang"—of which he speaks affectionately, as one might of an only child. The account of the great fight between Bhujang and the fancy of a certain Mr. Ahmed Khan, which took place one evening "after prayers," may be read by those who have a taste for such matters in Burton's book *Sind Revisited*.[1] When Bhujang died, Burton gave it almost Christian burial near his bungalow, and the facetious enquired whether the little mound was not "a baby's grave."

His hero was the eagle-faced little veteran and despot, Sir Charles Napier, generally known from his Jewish

(1) *Sind Revisited*. Vol. ii. pp. 78—83.

look as "Fagin," and from his irascibility as "The Devil's Brother," and after the war with Sind, the chief event of which was the battle of Meeanee (February 21st), where Sir Charles and Major Outram defeated the Ameer, his admiration grew almost to worship; though he did not actually see his hero till some months later. According to *Punch* the news of the battle was transmitted to headquarters in one word: "Peccavi." A quarrel then broke out between the great English leaders, and Western India was divided into the two opposing camps of Outramists and Napierists, Burton, of course, siding with the latter. In April, Burton returned to Bombay to present himself for examination in Hindustani, and having passed with honour[1] he returned to Baroda, where he experienced all the inconveniences attendant on the south-west monsoon. The rain fell in cataracts. Night and day he lay or sat in a wet skin; the air was alive with ants and other winged horrors, which settled on both food and drink, while the dust storms were so dense that candles had to be burned in mid-day. However he applied himself vigorously to Gujarati,[2] the language of the country, and also took lessons in Sanskrit.

"I soon," he says, "became as well acquainted as a stranger can with the practice of Hinduism. I carefully read up Ward, Moor, and the publications of the Asiatic Society . . . . and eventually my Hindu teacher officially allowed me to wear the Brahminical thread." He learnt some of the Hindu text books by heart,

(1) 5th May 1843. He was first of twelve.

(2) "How," asked Mr. J. F. Collingwood of him many years after, "do you manage to learn a language so rapidly and thoroughly?" To which he replied: "I stew the grammar down to a page which I carry in my pocket. Then when opportunity offers, or is made, I get hold of a native—preferably an old woman, and get her to talk to me. I follow her speech by ear and eye with the keenest attention, and repeat after her every word as nearly as possible, until I acquire the exact accent of the speaker and the true meaning of the words employed by her. I do not leave her before the lesson is learnt, and so on with others until my own speech is indistinguishable from that of the native."—Letter from Mr. Collingwood to me, 22nd June 1905.

including the Totá-kaháni,[1] which gave him a taste for "parrot books,"[2] on which he became an authority; while the study of the *Baital-Pachisi* led to his writing *Vikram and the Vampire.*[3] All this application caused his fellow officers to call him "The White Nigger."

Although, in after years, Burton often made bitter attacks on Christianity, and wrote most scathingly against the Roman Catholic priesthood, and the cenobitic life of the monks, yet at times he had certain sympathies with Roman Catholicism. Thus at Baroda, instead of attending the services of the garrison chaplain, he sat under the pleasant Goanese priest who preached to the camp servants; but he did not call himself a Catholic. In August he visited Bombay to be examined in Gujarati; and having passed with distinction, he once more returned to Baroda—just in time to join in the farewell revels of his regiment, which was ordered to Sind.

On board the Semiramis, in which the voyage was performed, he made the acquaintance of Captain Scott,

10. Karachi. Love of Disguise.

nephew of the novelist—a handsome man "with yellow hair and beard," and friendship followed. Both were fond of ancient history and romance, and Burton, who could speak Italian fluently and had knowledge of the canalization of the Po Valley, was able to render Scott, whose business was the surveyal of Sind, the precise assistance he just then required. Burton also formed a friendship with Dr. John Steinhäuser, afterwards surgeon at Aden. Then, too, it was at Karachi that he first saw his hero, Sir Charles

(1) The Totá-kaháni is an abridgment of the Tuti-námah (Parrot-book) of Nakhshabi. Portions of the latter were translated into English verse by J. Hoppner, 1805. See also *Anti-Jacobin Review* for 1805, p. 148.

(2) Unpublished letter to Mr. W. F. Kirby, 8th April 1885. See also Lib. Ed. of *The Arabian Nights*, viii., p. 73, and note to Night V.

(3) This book owes whatever charm it possesses chiefly to the apophthegms embedded in it. Thus, "Even the gods cannot resist a thoroughly obstinate man." "The fortune of a man who sits, sits also." "Reticence is but a habit. Practise it for a year, and you will find it harder to betray than to conceal your thoughts."

Napier. Though his ferocious temper repelled some, and his Rabelaisisms and kindred witticisms others, Sir Charles won the admiration and esteem of almost all who knew him. It was from him, to some extent, that Burton acquired the taste, afterwards so extraordinarily developed, for erotic, esoteric and other curious knowledge. Napier intensely hated the East India Company, as the champions of his detested rival, Major Outram, and customarily spoke of them contemptuously as the "Twenty-four kings of Leadenhall Street," while Burton on his part felt little respect for the effete and maundering body whose uniform he wore and whose pay he drew.

Karachi,(1) then not much better than a big village, was surrounded by walls which were perforated with "nostril holes," for pouring boiling water through in times of siege. There were narrow lanes, but no streets —the only open place being a miserable bazaar; while owing to the absence of sewers the stench was at times unendurable. Near the town was a great shallow artificial pond which abounded in huge sleepy crocodiles, sacred animals which were tended by a holy fakir, and one of Burton's amusements was to worry these creatures with his bull terrier. Tired of that pastime, he would muzzle a crocodile by means of a fowl fastened to a hook at the end of a rope, and then jump on to its back and take a zig-zag ride.(2) The feat of his friend, Lieutenant Beresford, of the 86th, however, was more daring even than that. Here and there in the pond were islets of rank grass, and one day noticing that the crocodiles and islets made a line across the pond, he took a run and hopped from one crocodile's back on to another or an islet until he reached the opposite side, though many a pair of huge jaws snapped angrily as he passed.

Burton presently found himself gazetted as Captain Scott's assistant; and having learnt the use of the theodolite and the spirit level, he went on December 10th

(1) Now it is a town of 80,000 inhabitants.

(2) *Sind Revisited*, i., 100

(1844) with a surveying party to Hyderabad[1] and the Guni River. The work was trying, but he varied it with hawking; and collected material for a work which he published eight years later with the title of *Falconry in the Valley of the Indus*. He then made the acquaintance of three natives, all of whom assisted him in his linguistic studies, Mirza Ali Akhbar,[2] Mirza Dàud, and Mirza Mohammad Musayn. Helped by the last he opened covertly at Karachi several shops with the object, however, not of making profit, but of obtaining intimate knowledge of the people and their secret customs. Then he put on long hair and a venerable beard, stained his limbs with henna, and called himself Abdullah of Bushire, a half-Arab. In this disguise, with spear in hand and pistols in holsters, he travelled the country with a little pack of nick-knacks. In order to display his stock he boldly entered private houses, for he found that if the master wanted to eject him, the mistress would be sure to oppose such a measure.

All his life he loved to disguise himself. We shall see him later as a Greek doctor, a Pathan Hakim, a dervise, an Arab pilgrim, a Moslem merchant, and an Arab shaykh. His shops had plenty of customers, for he was in the habit of giving the ladies, especially if they were pretty, "the heaviest possible weight for their money," though sometimes he would charge too much in order to induce them to chaffer with him. He learnt most, however, from the garrulity of a decayed beauty named Khanum Jan, who in her springtide had married a handsome tailor. Her husband having lost the graces of his person, she generally alluded to him affectionately as "that old hyena." This couple proved a Golconda for information. Burton had not long studied these and other persons before coming to the conclusion that the

(1) "The first City of Hind." See *Arabian Nights*, where it is called Al Mansurah, "Tale of Salim." Burton's A.N., Sup. i., 341. Lib. Ed. ix., 230.

(2) Mirza=Master. Burton met Ali Akhbar again in 1876. See chapter xviii., § 84.

Eastern mind is always in extremes, that it ignores what is meant by the "golden mean," and that it delights to range in flights limited only by the *ne plus ultra* of Nature herself. He picked up miscellaneous information about magic, white and black, Yoga [1], local manners and customs such as circumcision, both female and male, and other subjects, all of which he utilised when he came to write his Notes and Terminal Essay to *The Arabian Nights*, particularly the articles on Al Islam and woman. Then, too, when at Bombay and other large towns he used to ransack the bazaars for rare books and manuscripts, whether ancient or contemporaneous. Still, the most valuable portion of his knowledge was acquired orally.

11. A Dangerous Mission, 1845.

About this time it was reported to Sir Charles Napier that Karachi, though a town of only 2,000 souls, supported no fewer than three houses which were devoted to a particular and unspeakable vice [2] which is said to be common in the East. Sir Charles, whose custom it was to worm out the truth respecting anything and everything, at once looked round for someone willing to make enquiries and to report upon the subject. Burton being then the only British officer who could speak Sindi, the choice naturally fell upon him, and he undertook the task, only, however, on the express condition that his report should not be forwarded to the Bombay Government, from whom supporters of Napier's policy "could expect scant favour, mercy, or justice." Accompanied by his Munshi, Mirza Mohammed Hosayn Shiraz, and disguised as a merchant, Burton passed many evenings in the town, made the required visits, and obtained the fullest details, which were duly dispatched

(1) Yoga. One of the six systems of Brahmanical philosophy the essence of which is meditation. Its devotees believe that by certain ascetic practices they can acquire command over elementary matter. The Yogi go about India as fortune-tellers.

(2) Burton used to say that this vice is prevalent in a zone extending from the South of Spain through Persia to China and then opening out like a trumpet and embracing all aboriginal America. Within this zone he declared it to be endemic, outside it sporadic.

to Government House. But in 1847, when Napier quitted Sind "he left in his office Burton's unfortunate official." "This," says Burton, "found its way with sundry other reports to Bombay, and produced the expected result. A friend in the secretariat informed me that my summary dismissal had been formally proposed by one of Sir Charles Napier's successors, but this excess of outraged modesty was not allowed."(1) A little later, however, Burton had to suffer very severely for this unfortunate occurrence. Of course he heard regularly from home. His father was still immersed in blow-pipes and retorts, his mother still mildly protesting. His sister, who had won to herself for her loveliness the name of "the Moss Rose," was married to General Sir Henry Stisted,(2) his brother Edward was practising as an army doctor; his Grandmother Baker was dead.(3)

During one of his rambles he formed the acquaintance of a beautiful olive, oval-faced Persian girl of high descent·

12. **The Persian Beauty.**

We are told that her "eyes were narcissi, her cheeks sweet basil," her personal charms together with her siren voice and sweet disposition caused him to fall in love with her; but he had scarcely learnt that his passion was reciprocated before she died. We are told also that for many years he could never think of her without pain; and that when, some time after, he narrated the story to his sister he revealed considerable emotion. Miss Stisted thought she could see references to this episode in Burton's poem *The Kasîdah*, portions of which were written some three years later: "Mine eyes, my brain, my heart are sad—sad is the very core of me." This may be so, but the birth of a litter of pups, presented to him by his beloved bull terrier, seems to have taken the edge off his grief; and

(1) Burton's *Arabian Nights*, Terminal Essay, vol. x. pp. 205, 206, and *The Romance of Isabel Lady Burton*, by W. H. Wilkins, ii., 730.

(2) Married in 1845.

(3) She died 6th March 1846, aged 74.

his tribute to one of these pups, which received the name of Bachhûn, is really affecting.

The " Acting Commissioner " of the time was General Jacob of the Sind Horse, who wore a helmet of silver and a sabre-tache studded with diamonds. This, however, was not from pride or love of display, but because he held it policy in those who have to deal with Hindus not to neglect show and splendour. " In the eyes of Orientals," he used to remark, and Burton endorsed the saying, " no man is great unless he is also superbly dressed." As Jacob stuttered, one of his correspondents thought his name was J. J. J. J. J. Jacob, and terribly offended the testy General by writing it so. A brave and self-confident, but rancorous old man, Jacob by his senseless regulations brought the Indian army to the verge of ruin. This peccadillo was passed over, but a more serious offence, his inability to play whist, was remembered against him by his brother officers right to the day of his death. [1]

13. A Simian Dictionary.

When the Sikh war broke out Burton resigned his post under Scott in order to take part in the campaign in the Punjab, but peace being proclaimed a few weeks later, after the battle of Sobraon, Burton had no opportunities of distinguishing himself. So he returned to his studies, and now became ambitious to understand not only the people but also the monkeys of India. Consequently he collected some forty of them, made them live and eat after the manner of humans; and studied them as they mowed and gibbered. He would then talk to them and pronounce the sounds they made, until at last they could conduct quite a conversation together. Burton never divulged this talk, which, of course, may have been of a confidential nature, but he compiled a Simian Dictionary, and thus to some extent anticipated the work of Mr. R. L. Garner. Unfortunately the dictionary was some years later destroyed by fire.

(1) *He died 5th October, 1858. See Sind Revisited, ii. 261,*

We shall often notice in Burton's life what Burton himself called his dual nature. In the tale of Janshah in *The Arabian Nights* we read of a race of split men who separated longitudinally, each half hopping about contentedly on its own account, and reuniting with its fellow at pleasure. If Burton in a pre-existent state—and he half believed in the Pre-existence of Souls—belonged to this race, and one of his halves became accidentally united to one of the halves of somebody else, the condition of affairs would be explicable. In any circumstances, he was always insisting on his duality. For example—a kind-hearted man, who detested cruelty to animals, nevertheless he delighted, as we have seen, in the sport of cocking; an ambitious man, who wore himself out with his studies, yet he neutralised all his efforts to rise by giving way to an ungovernable temper. He would say just what he thought, and no man could have exhibited less tact. Thus he managed to give offence, and quite unnecessarily, to his superior officer, Colonel Henry Corsellis, and they were henceforth at handgrips.

14. Duality.

Among his favourite books was Jami's *Behâristan*. The only pity is that he did not take the advice proffered in the Third Garden:

"If Alexander's realm you want, to work adroitly go,
Make friends more friendly still, and make a friend of every foe."

Other instances of opposing qualities will be noticed as this work proceeds. Late in life, when he took to glasses, Burton used to say "My duality is proved by my eyes alone. My right eye requires a No. 50 convex lens, my left a No. 14." His assiduous application to his studies now brought about an illness, and, having returned to Bombay, he obtained two years' leave of absence to the salubrious Neilgherries.

CHAPTER IV

20TH FEBRUARY 1847—1849

UNDER THE SPELL OF CAMOENS

BIBLIOGRAPHY.

1. Grammar of the Jataki Dialect, 1849.
2. Remarks on Dr. Dorn's Christomathy of the Afghan Tong ue, 1849.
3. Reports on Sind addressed to the Bombay Government.
4. Grammar of the Mooltanee Language.

He left Goa on 20th February 1847, taking as usual a pattymar, his mind vibrant with thoughts of his great hero, the "Portingall" Camoens, with whose noble epic all Western India, from 15. Goa and Camoens.
Narsinga and Diu to Calicut is intimately associated. Passages from Camoens were frequently in his mouth, and in bitterest moments, in the times of profoundest dejection, he could always find relief in the pages of him whom he reverently calls "my master." Later in life he could see a parallel between the thorny and chequered career of Camoens and his own. Each spent his early manhood on the West Coast of India,(1) each did his country an incalculable service: Camoens by enriching Portugal with *The Lusiads*, Burton by his travels and by presenting to England vast stores of Oriental lore. Each received insult and ill-treatment, Camoens by imprisonment at Goa, Burton by the recall from Damascus. There was also a temperamental likeness between the two men. The passion for travel, the

(1) Camoens, born at Lisbon in 1524, reached Goa in 1553. In 1556 he was banished to Macao, where he commenced *The Lusiads*. He returned to Goa in 1558, was imprisoned there, and returned to Portugal in 1569. *The Lusiads* appeared in 1572. He died in poverty in 1580, aged 56.

love of poetry and adventure, the daring, the patriotism of Camoens all find their counterpart in his most painstaking English translator. Arrived at Panjim, Burton obtained lodgings and then set out by moonlight in a canoe for old Goa. The ruins of churches and monasteries fascinated him, but he grieved to find the once populous and opulent capital of Portuguese India absolutely a city of the dead. The historicity of the tale of Julnar the Sea Born and her son King Badr (1) seemed established, Queen Lab and her forbidding escort might have appeared at any moment. On all sides were bowing walls and tenantless houses. Poisonous plants covered the site of the Viceregal Palace, and monster bats hung by their heels at the corners of tombs. Thoughts of Camoens continued to impinge on his mind, and in imagination he saw his hero dungeoned and laid in iron writing his *Lusiads.* A visit to the tomb of S. Francis Xavier also deeply moved him. To pathos succeeded comedy. There was in Panjim an institution called the Caza da Misericordia, where young ladies, for the most part orphans, remained until they received suitable offers of marriage. The description of this place piqued Burton's curiosity, and hearing that it was not unusual for persons to propose themselves as suitors with a view to inspecting the curiosities of the establishment, he and some companions repaired to the Caza. Having seen the chapel and the other sights he mentioned that he wanted a wife. A very inquisitive duenna cross-examined him, and then he was allowed to interview one of the young ladies through a grating, while several persons, who refused to understand that they were not wanted, stood listening. Burton at once perceived that it would be an exhausting ordeal to make love in such circumstances, but he resolved to try, and a dialogue commenced as follows:

"Should you like to be married, senorita?"

"Yes, very much, señor."

"And why, if you would satisfy my curiosity?"

(1) *The Arabian Nights.*

"I don't know."

The rest of the conversation proved equally wooden and unsatisfactory, and quotations from poets were also wasted.

> "The maid, unused to flowers of eloquence
> Smiled at the words, but could not guess their sense."

Burton then informed the duenna that he thought he could get on better if he were allowed to go on the other side of the grating, and be left alone with the demure senorita. But at that the old lady suddenly became majestic. She informed him that before he could be admitted to so marked a privilege he would have to address an official letter to the mesa or board explaining his intentions, and requesting the desired permission. So Burton politely tendered his thanks, "scraped the ground thrice," departed with gravity, and in ten minutes forgot all about the belle behind the grille. It was while at Panjim, that, dissatisfied with the versions of Camoens by Strangford,[1] Mickle and others, Burton commenced a translation of his own, but it did not reach the press for thirty-three years.[2]

We next find him at Panany, whence he proceeded to Ootacamund, the sanitarium on the Neilgherries, where he devoted himself to the acquisition of Telugu, Toda, Persian and Arabic, though often interrupted by attacks of ophthalmia. While he was thus engaged, Sir Charles Napier returned to England (1847)[3] and Sind was placed under the Bombay Government "at that time the very sink of iniquity."[4]

In September Burton visited Calicut—the city above all others associated with Camoens, and here he had the pleasure of studying on the spot the scenes connected with the momentous landing of Da Gama as described in the

(1) Who was broken on the wheel by Lord Byron for dressing Camoens in "a suit of lace." See *English Bards and Scotch Reviewers*.

(2) Begun at Goa 1847, resumed at Fernando Po 1860-64, continued in Brazil and at Trieste. Finished at Cairo 1880.

(3) Napier was again in India in 1849. In 1851 he returned to England, where he died 29th August 1853, aged 71.

(4) *Life of Sir Charles Napier*, by Sir W. Napier.

seventh and most famous book of the *Lusiads*. In imagination, like Da Gama and his brave "Portingalls," he greeted the Moor Monzaida, interviewed the Zamorim, and circumvented the sinister designs of the sordid Catual; while his followers trafficked for strange webs and odoriferous gums. On his return to Bombay, reached on October 15th, Burton offered himself for examination in Persian, and gaining the first place, was presented by the Court of Directors with a thousand rupees. In the meantime his brother Edward, now more Greek-looking than ever, had risen to be Surgeon-Major, and had proceeded to Ceylon, where he was quartered with his regiment, the 37th.

16. "Would you a Sufi be?"

Upon his return to Sind, Burton at first applied himself sedulously to Sindi, and then, having conceived the idea of visiting Mecca, studied Moslem divinity, learnt much of the Koran by heart and made himself a "proficient at prayer." It would be unjust to regard this as mere acting. Truth to say, he was gradually becoming disillusioned. He was finding out in youth, or rather in early manhood, what it took Koheleth a lifetime to discover, namely, that "all is vanity." This being the state of his mind it is not suprising that he drifted into Sufism. He fasted, complied with the rules and performed all the exercises conscientiously. The idea of the height which he strove to attain, and the steps by which he mounted towards it, may be gathered from the Sufic poet Jami. Health, says Jami, is the best relish. A worshipper will never realise the pure love of the Lord unless he despises the whole world. Dalliance with women is a kind of mental derangement. Days are like pages in the book of life. You must record upon them only the best acts and memories.

"Would you a Sufi be, you must
Subdue your passions; banish lust
And anger; be of none afraid,
A hundred wounds take undismayed." (1)

(1) *The Behâristan*, 1st Garden.

In time, by dint of plain living, high thinking, and stifling generally the impulses of his nature, Burton became a Master Sufi, and all his life he sympathised with, and to some extent practised Sufism. Being prevented by the weakness of his eyes from continuing his survey work, he made a number of reports of the country and its people, which eventually drifted into print. Then came the stirring news that another campaign was imminent in Mooltan, his heart leaped with joy, and he begged to be allowed to accompany the force as interpreter. As he had passed examinations in six native languages and had studied others nobody was better qualified for the post or seemed to be more likely to get it.

17. Letter to Sarah Burton, 14th Nov. 1848.

It was while his fate thus hung in the balance that he wrote to his cousin Sarah[1] daughter of Dr. Francis Burton, who had just lost her mother.[2] His letter, which is headed Karachi, 14th November 1848, runs as follows:—"My dear cousin, I lose no time in replying to your note which conveyed to me the mournful tidings of our mutual loss. The letter took me quite by surprise. I was aware of my poor aunt's health having suffered, but never imagined that it was her last illness. You may be certain that I join with you in lamenting the event. Your mother had always been one of my best relations and kindest friends; indeed she was the only one with whom I kept up a constant correspondence during the last six years. I have every reason to regret her loss; and you, of course, much more. Your kind letter contained much matter of a consolatory nature; it was a melancholy satisfaction to hear that my excellent aunt's death-bed was such a peaceful one—a fit conclusion to so good and useful a life as hers was. You, too, must derive no small happiness from the reflection

(1) She married Col. T. Pryce Harrison. Her daughter is Mrs. Agg, of Cheltenham.

(2) She died 10th September 1848, and is buried at Elstree.

that both you and your sister[1] have always been dutiful daughters, and as such have contributed so much towards your departed mother's felicity in this life. In my father's last letter from Italy he alludes to the sad event, but wishes me not to mention it to my mother, adding that he has fears for her mind if it be abruptly alluded to.

At the distance of some 1,500[2] miles all we can do is to resign ourselves to calamities, and I confess to you that judging from the number of losses that our family has sustained during the last six years I fear that when able to return home I shall find no place capable of bearing that name. I hope, however, my dear cousin, that you or your sister will occasionally send me a line, informing me of your plans and movements, as I shall never leave to take the greatest interest in your proceedings. You may be certain that I shall never neglect to answer your letters and shall always look forward to them with the greatest pleasure. Stisted[3] is not yet out: his regiment is at Belgaum,[4] but I shall do my best to see him as soon as possible. Edward[5] is still in Ceylon and the war[6] has ceased there. I keep this letter open for ten or twelve days longer, as that time will decide my fate. A furious affair has broken out in Mooltan and the Punjaub and I have applied to the General commanding to go up with him on his personal staff. A few days more will decide the business—and I am not a little anxious about it, for though still suffering a little from my old complaint —ophthalmia—yet these opportunities are too far between to be lost."

Unfortunately for Burton, his official respecting his investigations at Karachi in 1845 was produced against

(1) Elisa married Colonel T. E. H. Pryce.

(2) That is from Italy, where his parents were living.

(3) Sir Henry Stisted, who in 1845 married Burton's sister.

(4) India, some 70 miles fron Goa.

(5) His brother. (6) The Ceylonese Rebellion of 1848.

Nov. 14. 1848.
Kurrachee

My dear Cousin / I lose no time in replying
to your note which conveyed to me the
mournful tidings of our mutual loss.
The letter took me quite by surprise. I was
aware of my poor Aunt's health having
suffered but never imagined that it was
her last illness. You may be certain that
I join with you in lamenting the event.
Your mother has always been one of my
best relations & kindest friends; indeed she
was the only one with whom I kept up a
constant correspondence during the last
six years. I have every reason to regret her
loss and you, of course have much
more

PORTION OF BURTON'S LETTER TO HIS COUSIN SARAH.
14TH NOVEMBER, 1848.

him,[1] and he was passed over[2] in favour of a man who knew but one language besides English. His theory that the most strenuous exertions lead to the most conspicuous successes now thoroughly broke down, and the scarlet and gold of his life, which had already become dulled, gave place to the "blackness of darkness." It was in the midst of this gloom and dejection that he wrote the postscript which he had promised to his cousin Sarah. The date is 25th November, 1848. He says "I am not going up to the siege of Mooltan, as the General with whom I had expected to be sent is recalled. Pray be kind enough to send on the enclosed to my father. I was afraid to direct it to him in Italy as it contains papers of some importance. You are welcome to the perusal, if you think it worth the trouble. I have also put in a short note for Aunt Georgiana. Kindly give my best love to your sister, and believe me, my dear cousin, your most affectionate R. Burton."

Chagrin and anger, combined with his old trouble, ophthalmia, had by this time sapped Burton's strength, a serious illness followed, and the world lost all interest for him.

18. Allahdad.

He returned to Bombay a complete wreck, with shrunken, tottering frame, sunken eyes, and a voice that had lost its sonority. "It is written," said his friends, "that your days are numbered, take our advice and go home to die." They carried him to his ship, "The Elisa," and as there seemed little hope even of his reaching England, he at once wrote a farewell letter to his mother. With him as servant, however, he had brought away a morose but attentive and good-hearted native named Allahdad, and thanks in part to Allahdad's good nursing, and in part to the bland and health-giving breezes of the ocean, he gradually regained his former health, strength, and vitality.

(1) See Chapter iii., § 11.

(2) See *Arabian Nights*, Terminal Essay § D, and *The Romance of Isabel Lady Burton*, vol. ii., p. 730.

At the time, he regarded these seven years spent in Sind as simply seven years wasted, and certainly his rewards were incommensurate with his exertions. Still, it was in Sind that the future became written on his forehead; in Sind that he began to collect that mass of amazing material which made possible his edition of *The Arabian Nights*.

CHAPTER V

1849 TO 3RD APRIL 1853

CHIEFLY BOULOGNE

BIBLIOGRAPHY :

5. Goa and the Blue Mountains, 1851.
6. Scinde ; or the Unhappy Valley, 2 vols., 1851.
7. Sindh, and the Races that Inhabit the Valley of the Indus, 1851.
8. Falconry in the Valley of the Indus, 1852.
9. Commencement with Dr. Steinhäuser of *The Arabian Nights*, 1852.
10. A complete System of Bayonet Exercise, 1853.

19. A Motto from Ariosto.

When " The Elisa " approached Plymouth, with its " turfy hills, wooded parks and pretty seats," Allahdad opened his eyes in wonderment. " What manner of men must you English be," he said, " to leave such a paradise and travel to such a pandemonium as ours without compulsion ? " On arriving in London, Burton called on his Aunt Georgiana,[1] flirted with his pretty cousins Sarah and Elisa, attended to business of various kinds, and then, in company with Allahdad, set out for Italy to see his father and mother, who were still wandering aimlessly about Europe, and inhaling now the breath of vineyard and garden and now the odours of the laboratory. He found them, his sister, and her two little daughters, Georgiana and Maria (Minnie) at Pisa, and the meeting was a very happy one. Burton's deep affection for his parents, his sister and his brother, is forced upon our notice at every turn ; and later he came to regard his

(1) His Grandmother Baker had died in 1846.

nieces just as tenderly. Quoting Coleridge, he used to say:

"To be beloved is all I need,
And whom I love I love indeed." [1]

If Burton was thus drawn to those nearest of kin to him, so also his warm heart welled with affection for his friends, and for those who did him kindnesses. "If you value a man or his work," he said, "don't conceal your feelings." The warmth of his affection for his friends Drake, Arbuthnot, and others, will be noticed as this book proceeds. On one occasion, after a spontaneous outburst of appreciation, he said in palliation of his enthusiasm, "Pardon me, but this is an asthenic age—and true-hearted men are rare." Presently we find him revisiting some of his old haunts. In his youth he had explored Italy almost from end to end; but the literary associations of the various towns were their principal charm. To him, Verona stood for Catullus, Brindisi for Virgil, Sorrento for Tasso, Florence for "the all Etruscan three,"[2] Dante, Petrarch, and Boccaccio, Reggio and Ferrara for Ariosto. It was from Ariosto, perhaps through Camoens, who adopted it, that he took his life motto, "Honour, not honours"—

"'Tis honour, lovely lady, that calls me to the field,
And not a painted eagle upon a painted shield." [3]

All the Burton servants obtained some knowledge ot Italian, even Allahdad being soon able to swear fluently in it, and his aptitude, joined to a quarrelsome temper and an illogical prejudice against all Italians, caused innumerable broils.

By and by the family returned to England and Miss Stisted thus describes the progress: "One of the earliest pictures in my memory is of a travelling carriage crossing snow-covered Alps. A carriage containing my mother

(1) The Pains of Sleep.

(2) Byron: *Childe Harold*, iv. 56.

(3) Ariosto's *Orlando* was published in 1516; *The Lusiads* appeared in 1572.

and uncle, sister and self, an English maid, and a romantic but surly Asiatic named Allahdad. Richard Burton, handsome, tall and broad-shouldered, was oftener outside the carriage than in it, as the noise made by his two small nieces rendered pedestrian exercise, even in the snow, an agreeable and almost necessary variety." Now and then he gave them bits of snow to taste, which they hoped might be sugar.[1] On reaching England he sent Allahdad back to Bombay.

Much of the year 1850 was spent at Leamington and Dover, and in 1851, Burton, accompanied by his brother Edward, crossed to Boulogne, where he prepared for publication his books, *Goa, Scinde, Falconry in the Valley of the Indus*, and *Bayonet Exercise.* Love of a sort mingled with literature, for he continued various flirtations, but without any thought of marriage; for he was still only a lieutenant in the service of John Company, and his prospects were not rosy. We said "love of a sort," and advisedly, for we cannot bring ourselves to believe that Burton was ever frenziedly in love with any woman. He was, to use his own expression, no "hot amorist." Of his views on polygamy, to which he had distinct leanings, we shall speak later. He said he required two, and only two qualities in a woman, namely beauty and affection. It was the Eastern idea. The Hindu Angelina might be vacuous, vain, papilionaceous, silly, or even a mere doll, but if her hair hung down "like the tail of a Tartary cow,"[2] if her eyes were "like the stones of unripe mangoes," and her nose resembled the beak of a parrot, the Hindu Edwin was more than satisfied. Dr. Johnson's "unidead girl" would have done as well as the blue-stocking Tawaddud.[3]

(1) *Temple Bar*, vol. xcii., p. 335.

(2) As did that of the beauty in *The Baital-Pachisi—Vikram and the Vampire.* Meml. Ed., p. 228.

(3) Tale of Abu-el-Husn and his slave girl, Tawaddud.—*The Arabian Nights*,

20. Isabel Arundell & "My Dear Louisa." 1851.

It was during Burton's stay at Boulogne that he saw the handsome girl who ten years later became his wife—Isabel, daughter of Mr. Henry Raymond Arundell. She was the eldest of a very large family. Just twenty, fair, "with yards of golden hair," dark blue eyes and a queenly manner, Isabel Arundell everywhere attracted attention. No portrait, it was said, ever did justice to her virginal beauty. "When she was in any company you could look at no one else," the charm of her manner exceeded even the graces of her person, but her education was defective, and she was amusingly superstitious. She could be heard saying at every turn: "This is a good omen; that a bad one; oh, shocking! the spoons are crossed;

> By the pricking of my thumbs
> Something wicked this way comes."

Though not themselves wealthy, the Arundells were of noble lineage, and had rich and influential relations who prided themselves on being "old English Catholics." Among Miss Arundell's ancestors was Henry, 6th Lord Arundell of Wardour; her grandfather and the 9th Lord were brothers; and her mother was sister to Lord Gerard.

Isabel Arundell and Burton could have conducted their first conversation just as well had they been deaf and dumb. Strolling on the ramparts he noticed a bevy of handsome girls, one of whom, owing to her exceptional looks, particularly fired him, and having managed to attract her attention, he chalked on a wall, "May I speak to you," and left the piece of chalk at the end of the sentence. She took it up and wrote under it, "No, mother will be angry."

She had, however, long pictured to herself an ideal husband, and on seeing Burton, she exclaimed under her breath: "That is the man!" She describes him as "five feet eleven inches in height, very broad, thin and muscular, with very dark hair, black, clearly defined, sagacious

RICHARD BURTON AND HIS SISTER, LADY STISTED

FROM AN OIL PAINTING BY JACQUAND, DONE ABOUT 1851

THE ORIGINAL IS IN THE POSSESSION OF MR. E. S. MOSTYN PRYCE, OF GUNLEY HALL, STAFFORDSHIRE.

Photo by J. H. Anderson and Son, Welshpool.

eyebrows, a brown, weather-beaten complexion, straight Arab features, a determined-looking mouth and chin, nearly covered by an enormous moustache; two large, black, flashing eyes, with long lashes," and a "fierce, proud, melancholy expression." (1) In the words of one of his friends, he had the eye of an angel, the jaw of a devil. Also staying at Boulogne was a young lady for whom Burton entertained a sincere affection, and whom he would probably have married but for the poorness of his outlook. "My dear Louisa," (2) as he called her, was a relative of Miss Arundell, and hearing what had occurred, she did Burton and Miss Arundell the kindness of formally introducing them to each other. Miss Arundell never tried to attract Burton's attention—we have her word for that—but wherever he went she went too; and she never lost an opportunity of accidentally crossing his path. She glanced furtively at him, and read, with mantling cheeks, some lines of poetry which he sent her. She considered sacred a sash which she wore when dancing with him, and she remembered him specially in her prayers. Henceforward, one devouring desire occupied her mind. She wished—and praiseworthily—to be Burton's wife. To him, on the other hand, she was but an ephemeral fancy—one of the hundred and fifty women—his fair cousins in England and the softer and darker beauties of France and Italy—to whom he had said tender nothings. Later, when Miss Arundell saw him flirting with another girl, a certain "Louise" (3) (not to be confused with "my dear Louisa"), she bridled up, coloured to her brow-locks, called "Louise" "fast" and Louise's mother "vulgar." Naturally they would be. (4) With "myosotis eyes," peachy cheeks and auburn hair, rolling over

(1) *Life*, i., 167.

(2) She became Mrs. Segrave.

(3) See Burton's *Stone Talk*, 1865. Probably not "Louise" at all, the name being used to suit the rhyme.

(4) Mrs. Burton was always very severe on her own sex.

ivory shoulders,(1) "Louise" was progressing admirably, when, unfortunately for her, there came in view a fleshy, vinous matron of elephantine proportions, whom she addressed as "mother." The sight of this caricature of the "Thing Divine," to use Burton's expression, and the thought that to this the "Thing Divine" would some day come, instantly quenched his fires, and when the mother tried to bring him to a decision, by inquiring his intentions regarding her daughter, he horrified her by replying: "Strictly dishonourable, madam." "Englishmen," he reflected, "who are restricted to one wife, cannot be too careful." Miss Arundell was also jealous of "My dear Louisa," though unwarrantably, for that lady presently became Mrs. Segrave; but she and Burton long preserved for each other a reminiscitory attachment, and we shall get several more glimpses of her as this book proceeds.(2)

Isabel Arundell was herself somewhat cheered by the prophecy of a gipsy of her acquaintance—one Hagar Burton—who with couched eyes and solemn voice not only prognosticated darkly her whole career, but persistently declared that the romance would end in marriage; still, she fretted a good deal, and at last, as persons in love sometimes do, became seriously indisposed. Without loss of time her parents called in a skilful physician, who, with his experienced eye, saw at once that it was indigestion, and prescribed accordingly. Residing at Boulogne in 1851, was a French painter named François Jacquand, who had obtained distinction by his pictures of monks, and "a large historical tableau representing the death chamber of the Duc d'Orleans." In an oil painting which he made of Burton and his sister, and which is here reproduced for the first time, Burton appears as a pallid young military man, heavily moustached, with large brown eyes;(3) and his worn and somewhat

(1) See *Stone Talk*.

(2) See Chapter x.

(3) The original, which belonged to Miss Stisted, is now in the possession of Mr. Mostyn Pryce, of Gunley Hall.

MISS GEORGIANA STISTED.

MISS MARIA STISTED.

FROM OIL PAINTINGS IN THE POSSESSION OF MR. E. S. MOSTYN PRYCE, OF GUNLEY HALL, STAFFORDSHIRE.

Photos by J. H. Anderson and Son, Welshpool.

melancholy face is a striking contrast to the bright and cheerful looks of his comely sister. Our portraits of the Misses Stisted are also from paintings by Jacquand. Burton's habit of concealing his ailments which we noticed as a feature of his boyhood was as conspicuous in later life. "On one occasion," says Miss Stisted, "when seized with inflammation of the bladder, a fact he tried to keep to himself, he continued to joke and laugh as much as usual, and went on with his reading and writing as if little were the matter. At last the agony became too atrocious, and he remarked in a fit of absence 'If I don't get better before night, I shall be an angel.' Questions followed, consternation reigned around, and the doctor was instantly summoned."

When Burton first became acquainted with Forster FitzGerald Arbuthnot is uncertain; but by 1853, they were on terms of intimacy. Burton was then 32, Arbuthnot 20. Of this enormously important fact in Burton's life—his friendship with Arbuthnot—no previous writer has said a single word, except Lady Burton, and she dismisses the matter with a few careless sentences, though admitting that Arbuthnot was her husband's most intimate friend. Of the strength of the bond that united the two men, and the admiration felt by Arbuthnot for Burton, she had little idea. F. F. Arbuthnot, born in 1833, was second son of Sir Robert Keith Arbuthnot and Anne, daughter of Field-Marshal Sir John Forster FitzGerald, G.C.B. Educated at Haileybury, he entered in 1852 the Bombay Civil Service, and rose subsequently to the important position of "collector." A man of a quiet and amiable disposition, Arbuthnot never said an unkind word either to or about anyone. The sweetness and serenity of his manner were commented upon by all his friends; but like so many of your quiet men, he had a determination—a steady heroism, which made everything give way. Oppose Burton, and you would instantly receive a blow aimed straight from the shoulder,

21. Forster FitzGerald Arbuthnot 1853.

oppose Arbuthnot and you would be pushed quietly and amiably aside—but pushed aside nevertheless. A great idea had early possessed him. He wanted to see as much attention paid to the literatures of India, Persia and Arabia as to those of ancient Greece and Rome. All the famous books of the East, he said, should be translated into English—even the erotic, and he insisted that if proper precautions were taken so that none but scholars could obtain them, no possible harm could ensue.[1]

"England," he wrote long after (1887), "has greater interests in the East than any other country in Europe, and ought to lead the way in keeping the world informed on all subjects connected with Oriental literature. Surely the time has not yet arrived for her to take a back seat on that coach, and to let other nations do a work which she ought to do herself."[2] The expression "on that coach," by the by, was eminently characteristic of a man who plumed himself on being a Jehu of Jehus. Hundreds of invaluable manuscripts written by poets and sages, he said, require to be translated into English, and the need of the day is an Oriental Translation Fund. A man of means, Arbuthnot was sometime later to apply his money to the cause he had at heart; and year in, year out, we shall find him and Burton striking at the self-same anvil. Though there was a considerable difference in their ages, and though thousands of miles often separated them, their minds were ever united, and they went down the stream of life together like two brothers.

(1) Of course, since Arbuthnot's time scores of men have taken the burden on their shoulders, and translations of the Maha-Bharata, the Ramayana, and the works of Kalidasa, Hafiz, Sadi and Jami, are now in the hands of everybody.

(2) Preface to *Persian Portraits*.

CHAPTER VI

3RD APRIL 1853 TO 29TH OCTOBER 1854

PILGRIMAGE TO MECCA

BIBLIOGRAPHY:

11. The Kasîdah (commenced).
12. El Islam (commenced).

22. The Man Wants to Wander.

Much of his time at Boulogne Burton devoted to fencing; and to his instructor, M. Constantin, he paid glowing tributes. He thoroughly mastered the art, defeated all antagonists, whether English or French earned his "*brevet de pointe* for the excellence of his swordsmanship, and became a Maître d' Armes." As horseman, swordsman, and marksman, no soldier of his day surpassed him, and very few equalled him. But of fencing, flirting and book-writing, he soon got heartily tired. Like his putative ancestors, the gipsies, he could never be happy long in one place. He says, "The thorough-bred wanderer's idiosyncrasy, I presume to be a composition of what phrenologists call inhabitiveness and locality equally and largely developed. After a long and toilsome march, weary of the way, he drops into the nearest place of rest to become the most domestic of men. For a while he smokes the pipe of permanence with an infinite zest, he delights in various siestas during the day, relishing withal a long sleep at night; he enjoys dining at a fixed dinner hour, and wonders at the demoralisation of the mind which cannot find means of excitement in chit-chat or small talk, in a novel or a newspaper. But soon the passive fit has passed away; again a paroxysm of ennui coming on by slow degrees, viator loses appetite, he walks about his room all night, he yawns at conversations, and a book acts upon him as a narcotic. The man wants to wander, and he must do so, or he shall die." (1)

As we have seen, Burton, even before he left Sind, had burned to visit Mecca. Four years had since elapsed, and his eyes still turned towards "Allah's holy house." Having obtained another twelve months' furlough, in order that he "might pursue his Arabic studies in lands where the language is best learned," he formed the bold plan of crossing Arabia from Mecca to the Persian Gulf. Ultimately, however, he decided, in emulation of Burckhardt, the great traveller, to visit Medina and Mecca in the disguise of a pilgrim, a feat that only the most temerarious of men would have dared even to dream of. He made every conceivable preparation, learning among other usefulnesses how to forge horse shoes and to shoe a horse. To his parents and Lady Stisted and her daughters, who were then residing at Bath, he paid several visits, but when he last parted from them with his usual "Adieu, sans adieu," it did not occur to them that he was about to leave for good; for he could not—he never could—muster up sufficient courage to say a final "Good-bye." Shortly after his departure his mother found a letter addressed to her and in his handwriting. It contained, besides an outline of his dangerous plans, the instruction that, in case he should be killed, his "small stock of valuables" was to be divided between her and his sister.

23. Haji Wali, 1853.

Once more Burton had the keen pleasure of putting on disguise. Richard F. Burton ceased to be, and a muscular and powerful Mirza Abdullah, of Bushire, took his place. "I have always wished to see," he explained to a friend, "what others have been content to hear of." He wore long hair and Oriental costume, and his face and limbs

(1) *Pilgrimage to El-Medinah and Meccah,* Memorial Ed., vol. i., p. 16.

were stained with henna. Accompanied by Captain Henry Grindlay of the Bengal Cavalry, he left London for Southampton, 3rd April 1853, and thence took steamer for Egypt, without ever a thought of Isabel Arundell's blue eye or Rapunzel hair, and utterly unconscious of the sighs he had evoked. At Alexandria he was the guest of Mr. John Thurburn and his son-in-law, Mr. John Larking,[1] at their residence "The Sycamores," but he slept in an outhouse in order the better to delude the servants. He read the Koran sedulously, howled his prayers with a local shaykh who imparted to him the niceties of the faith, purified himself, made an ostentatious display of piety, and gave out that he was a hakim or doctor preparing to be a dervish. As he had some knowledge of medicine, this rôle was an easy one, and his keen sense of humour made the experience enjoyable enough. On the steamer that carried him to Cairo, he fraternized with two of his fellow-passengers, a Hindu named Khudabakhsh and an Alexandrian merchant named Haji Wali. Haji Wali, whose connection with Burton lasted some thirty years,[2] was a middle-aged man with a large round head closely shaven, a bull neck, a thin red beard, handsome features which beamed with benevolence, and a reputation for wiliness and cupidity. Upon their arrival at Boulak, the port of Cairo, Khudabakhsh, who lived there, invited Burton to stay with him. Hindu-like, Khudabakhsh wanted his guest to sit, talk, smoke, and sip sherbet all day. But this Burton could not endure. Nothing, as he says, suits the English less than perpetual society, "an utter want of solitude, when one cannot retire into one self an instant without being asked some puerile question by a companion, or look into a book without a servant peering over one's shoulder." At last, losing all patience,

(1) Burton dedicated to Mr. John Larking the 7th volume of *The Arabian Nights*.

(2) Haji Wali in 1877 accompanied Burton to Midian. He died 3rd August 1883, aged 84. See Chapter xx.

he left his host and went to a khan, where he once more met Haji Wali. They smoked together the forbidden weed hashish, and grew confidential. Following Haji Wali's advice, Burton, having changed his dress, now posed as an Afghan doctor, and by giving his patients plenty for their money and by prescribing rough measures which acted beneficially upon their imaginations, he gained a coveted reputation. He always commenced his prescriptions piously with: "In the name of Allah, the compassionate, the merciful, and blessings and peace be upon our Lord the Apostle"; and Haji Wali vaunted him as "the very phœnix of physicians." According to his wont, he never lost an opportunity of learning the ways and customs of the various people among whom he was thrown, or of foisting himself on any company in which he thought he could increase his knowledge. His whole life indeed was a preparation for "*The Arabian Nights.*" Thus at Cairo he had the good fortune to cure some Abyssinian slave-girls of various complaints, including the "price-lowering habit of snoring," and in return he made the slave dealer take him about the town and unfold the mysteries of his craft. He also visited the resting-place of his hero, Burckhardt;[1] indeed, in whatever town he sojourned, he sought out the places associated with the illustrious dead. It was now the Ramazan, and he observed it by fasting, reading the Koran, and saying countless prayers with his face turned devoutly to the Kiblah.[2] He heartily rejoiced, however, with the multitude when the dreary month was over, and he describes[3] amusingly the scenes on the first day following it: "Most people," he says, "were in fresh suits of finery; and so strong is personal vanity in the breasts of Orientals . . that from Cairo to Calcutta it would be difficult to find a sad heart under a handsome coat. The men swaggered, the women minced their

(1) He died at Cairo, 15th October 1817.

(2) That is, in the direction of Mecca.

(3) *Pilgrimage*, Memorial Ed., i., 116.

steps, rolled their eyes, and were eternally arranging, and coquetting with their head-veils.'' In the house of a friend he saw an Armenian wedding. For servant he now took a cowardly and thievish lad named Nur, and, subsequently, he made the acquaintance of a Meccan youth, Mohammed, who was to become his companion throughout the pilgrimage. Mohammed was 18, chocolate brown, short, obese, hypocritical, cowardly, astute, selfish and affectionate. Burton not only purchased the ordinary pilgrim garb, but he also took the precaution to attach to his person "a star sapphire,'' the sight of which inspired his companions with ''an almost reverential awe,'' and even led them to ascribe to him thaumaturgic power.[1] His further preparations for the sacred pilgrimage reads rather like a page out of Charles Lever, for the rollicking Irishman was as much in evidence as the holy devotee. They culminated in a drinking bout with an Albanian captain, whom he left, so to speak, under the table; and this having got noised abroad, Burton, with his reputation for sanctity forfeited, found it expedient to set off at once for Mecca. He sent the boy Nur on to Suez with his baggage and followed him soon after on a camel through a ''haggard land infested with wild beasts and wilder men.'' At Suez he made the acquaintance of some Medina and Mecca folk, who were to be his fellow-travellers; including ''Sa'ad the Demon,'' a negro who had two boxes of handsome apparel for his three Medina wives and was resolved to ''travel free;'' and Shaykh Hamid, a ''lank Arab foul with sweat,'' who never said his prayers because of the trouble of taking clean clothes out of his box. ''All these persons,'' says Burton, ''lost no time in opening the question of a loan. It was a lesson in Oriental metaphysics to see their condition. They had a twelve days' voyage and a four days' journey before them; boxes to carry, custom houses to face, and stomachs to fill; yet the whole party could scarcely,

(1) See Preface to *The Kasîdah*, Edition published in 1894.

I believe, muster two dollars of ready money. Their boxes were full of valuables, arms, clothes, pipes, slippers, sweetmeats, and other 'notions,' but nothing short of starvation would have induced them to pledge the smallest article."[1] Foreseeing the advantage of their company, Burton sagaciously lent each of them a little money at high interest, not for the sake of profit, but with a view to becoming a Hatim Tai,[2] by a "never mind" on settling day. This piece of policy made "the Father of Moustaches," as they called him, a person of importance among them. During the delay before starting, he employed himself first in doctoring, and then in flirting with a party of Egyptian women the most seductive of whom was one Fattumah,[3] a plump lady of thirty "fond of flattery and possessing, like all her people, a voluble tongue." The refrain of every conversation was "Marry me, O Fattumah! O daughter! O female pilgrim." To which the lady would reply coquettishly, "with a toss of the head and a flirting manipulation of her head veil," "I am mated, O young man." Sometimes he imitated her Egyptian accent and deprecated her country women, causing her to get angry and bid him begone. Then, instead of "marry me, O Fattumah," he would say, "O old woman and decrepit, fit only to carry wood to market." This would bring a torrent of angry words, but when they met again all was forforgotten and the flirtations of the day before were repeated.

Burton and his party now embarked on the sambuk which was to take them to Yambu, the port of Medina.

24. The Pilgrim Ship. 6th July 1853.

As ninety-seven pilgrims were crowded on a vessel constructed to carry only sixty, most extraordinary scenes occurred. Thanks to the exertions of Sa'ad the Demon, Burton and his friends secured places on the poop, the

(1) *Pilgrimage*, Memorial Ed., i., 165.

(2) A chieftain celebrated for his generosity. There are several stories about him in *The Arabian Nights*.

(3) An incrementative of Fatimah.

most eligible part of the vessel. They would not be very comfortable anywhere, Sa'ad explained, but "Allah makes all things easy." Sa'ad himself, who was blessed with a doggedness that always succeeds, managed to get his passage free by declaring himself an able seaman. Disturbances soon commenced. The chief offenders were some Maghrabis, "fine looking animals from the deserts about Tripoli" the leader of whom, one Maula Ali, "a burly savage," struck Burton as ridiculously like his old Richmond schoolmaster, the Rev. Charles Delafosse. These gentry tried to force their way on to the poop, but Sa'ad distributed among his party a number of ash staves six feet long, and thick as a man's wrist. "He shouted to us," says Burton, "'Defend yourselves if you don't wish to be the meat of the Maghrabis!' and to the enemy 'Dogs and sons of dogs! now shall you see what the children of the Arab are.' 'I am Omar of Daghistan!' 'I am Abdullah the son of Joseph!' 'I am Sa'ad the Demon! (1)' we exclaimed." And, Burton, with his turbulent blood well stirred, found himself in the seventh heaven. "To do our enemies justice," he continues, "they showed no sign of flinching; they swarmed towards the poop like angry hornets, and encouraged each other with cries of 'Allaho Akbar!' But we had a vantage ground about four feet above them, and their short daggers could do nothing against our terrible quarter staves. Presently a thought struck me. A large earthen jar full of drinking water, in its heavy frame of wood stood upon the edge of the poop. Seeing an opportunity, I crept up to the jar and rolled it down upon the swarm of assailants. Its fall caused a shriller shriek to rise above the ordinary din, for heads, limbs and bodies were sorely bruised by the weight, scratched

(1) Burton says of the Arabs, "Above all their qualities, personal conceit is remarkable; they show it in their strut, in their looks, and almost in every word. 'I am such a one, the son of such a one,' is a common expletive, especially in times of danger; and this spirit is not wholly to be condemned, as it certainly acts as an incentive to gallant actions."—*Pilgrimage*, ii, 21., Memorial Ed.

by the broken potsherds, and wetted by the sudden discharge.''[1] The Maghrabis then slunk off towards the end of the vessel, and presently solicited peace.''

The beauties of the sunrise baffled description. The vessel sailed over a violet sea, and under a sky dappled with agate-coloured clouds. At noon the heat was terrible and all colour melted away, '' with the canescence from above.'' The passengers were sympathetic with one another, notwithstanding their recent factiousness, and were especially kind to a poor little brown baby, which they handed round and nursed by turns, but the heat, the filth, and the stench of the ship defied description. At Mahar, one of the places where they landed, Burton injured his foot with a poisonous thorn, which made him lame for the rest of the pilgrimage. Presently the welcome profile of Radhwa came in view, the mountain of which the unfortunate Antar[2] sang so plaintively :

> " Did Radhwa strive to support my woes,
> Radhwa itself would be crushed by the weight,"

and on July 17th, after twelve days of purgatory, Burton sprang on shore at Yambu.

He now dressed himself as an Arab, that is to say, he covered his head with a red kerchief bordered with yellow, his body with a cotton shirt and a camel's hair cloak, while a red sash, a spear and a dagger completed the outfit. Then, having hired some camels, he joined a caravan, consisting of several hundred men and beasts, which was bound for Medina ; but his injured foot still incommoded him. Determined, however, to allow nobody to exceed him in piety, he thrice a day or oftener pounded the sand with his forehead like a true Mussulman.

25. Medina.

While passing through one of the mountain gorges the pilgrims were attacked by a number of predatory Bedouin,

(1) *Pilgrimage to Meccah*, Memorial Ed., i., 193.

(2) A creation of the poet Al-Asma'i. He is mentioned in *The Arabian Nights*.

led by a ferocious chief named Saad, who fired upon them from the rocks with deadly effect, but, at last, after a journey of 130 miles, they reached Medina, with the great sun-scorched Mount Ohod towering behind it—the holy city where, according to repute, the coffin of Mohammed swung between heaven and earth.[1] Medina consisted of three parts, a walled town, a large suburb, with ruinous defences, and a fort. Minarets shot up above the numerous flat roofs, and above all flashed the pride of the city, the green dome that covered the tomb of Mohammed. Burton became the guest of the dilatory and dirty Shaykh Hamid. The children of the household, he says, ran about in a half nude state, but he never once set eyes upon the face of woman, "unless the African slave girls be allowed the title. Even these at first attempted to draw their ragged veils over their sable charms." Having dressed themselves in white, Burton and Hamid sallied out for the Prophet's Tomb, Burton riding on a donkey because of his lameness. He found the approach to the Mosque choked up by ignoble buildings, and declares that as a whole it had neither beauty nor dignity. Upon entering, he was also disillusioned, for its interior was both mean and tawdry. After various prayers they visited first the "Hujrah," where they saw the tombs of Mohammed, Abu Bakr Omar and Fatimah; and afterwards El Rauzah, the Garden situated between the Hujrah and the Prophet's Pulpit, both very celebrated spots. Of the latter, Mohammed said: "Between my house and my pulpit is a garden of the gardens of paradise."[2] After more prayers they wandered round to the other sights, including the fine Gate of Salvation, the five minarets, and the three celebrated pillars, called respectively, Al-Mukhallak,

(1) How this tradition arose nobody seems to know. There are several theories.

(2) It is decorated to resemble a garden. There are man references to it in the *Arabian Nights*. Thus the tale of Otbah and Rayya (Lib. Ed., v., 289) begins "One night as I sat in the garden between the tomb and the pulpit."

the Pillar of Ayishah, and the Pillar of Repentance. They then made their way to the Mosque of Kuba, some two miles out of the town, and witnessed the entry into Medina of the great caravan from Damascus, numbering 7,000 souls—grandees in gorgeous litters of green and gold, huge white Syrian dromedaries, richly caparisoned horses and mules, devout Hajis, sherbet sellers, water carriers, and a multitude of camels, sheep and goats.(1) Lastly Burton and his friends pilgrimaged to the holy Mount Ohod with its graves of "the martyrs;" and to the celebrated Al-Bakia, or Saints' Cemetery, where lie ten thousand of the Prophet's companions. On entering the latter they repeated the usual salutation: "Peace be upon ye, O People of Al-Bakia, and then sought out the principal tombs—namely those of the Caliph Othman,(2) "Our Lady Halimah,"(3) the Infant Ibrahim,(4) and about fourteen of Mohammed's wives.(5) The cemetery swarmed with clamorous beggars, who squatted with dirty cotton napkins spread on the ground before them for the reception of coins. Some of the women promised to recite Fatihahs for the donors, and the most audacious seized the visitors by their skirts. Burton laid out three dollars in this way, but though the recipients promised loudly to supplicate Allah in behalf of his lame foot, it did not perceptibly benefit. Burton's companions hinted that he might do worse than settle in Medina. "Why not," said one, "open a shop somewhere near the Prophet's Mosque? There thou wilt eat bread by thy skill, and thy soul will have the blessing of being on holy ground." Burton, however, wanted to be going forward.

On 31st August, after praying "a two-bow prayer," he bade adieu to Shaykh Hamid, and with Nur and the

(1) *Pilgrimage to Meccah* (Mem. Ed., i., 418).

(2) Mohammed's son-in-law. (3) Mohammed's wet nurse. (4) Son of Mohammed and the Coptic girl Mariyah, sent to Mohammed as a present by Jarih, the Governor of Alexandria. ( ) Khadijah, the first wife, lies at Mecca.

BURTON IN THE DESERT.

FROM THE PAINTING IN THE CENTRAL LIBRARY, CAMBERWELL

(By permission of the Committee).

boy Mohammed, joined the caravan bound for Mecca, the route taken being the celebrated road through the arid Nejd made by Zubaydah, wife of Harun al Rashid. The events of the journey 26. Mecca. were not remarkable, though Mohammed very nearly killed himself by feeding too liberally on clarified butter and dates mashed with flour. Sometimes Burton cheered the way and delighted his companions by singing the song of Maysunah, the Arab girl who longed to get back from the Caliph's palace to the black tents of her tribe. Everybody got into good humour when he began :

> "Oh take these purple robes away,
> Give back my cloak of camel's hair,"

and they laughed till they fell on their backs when he came to the line where the desert beauty calls her Royal husband a "fatted ass." In truth, they needed something to cheer them, for the sky was burnished brass, and their goats died like flies. Simoon and sand-pillar threw down the camels, and loathsome vultures ready for either beast or man hovered above or squabbled around them. To crown their discomforts they were again attacked by the Bedouin, whom they dispersed only after a stubborn fight and with the loss of several dromedaries. After passing the classic Wady Laymun, sung by the Arab poet Labid[1] in lines suggestive of Goldsmith's *Deserted Village*, they very piously shaved their heads and donned the conventional attire, namely two new cotton cloths with narrow red stripes and fringes ; and when the Holy City came in view, the whole caravan raised the cry, "Mecca ! Mecca ! the Sanctuary ! O the Sanctuary ! Labbayk ! Labbayk !"[2] the voices being not infrequently broken by sobs.

On entering the gates, Burton and Nur crossed the famous hill Safa and took up their abode with the lad

[1] Known to us chiefly through Dr. Carlyle's poor translation. See *Pilgrimage*, ii., 147

[2] Here am I.

Mohammed. Early next morning they rose, bathed, and made their way with the crowd to the Prophet's Mosque in order to worship at the huge bier-like erection called the Kaaba, and the adjacent semi-circular Hatim's wall. The famous Kaaba, which is in the middle of the great court-yard, looked at a distance like an enormous cube, covered with a black curtain, but its plan is really trapeziform. "There at last it lay," cries Burton, "the bourn of my long and weary pilgrimage, realising the plans and hopes of many and many a year,"—the Kaaba, the place of answered prayer, above which in the heaven of heavens Allah himself sits and draws his pen through people's sins. "The mirage of fancy invested the huge catafalque and its gloomy pall with peculiar charms." Of all the worshippers who clung weeping to the curtain,[1] or who pressed their beating hearts to the sacred black stone built into the Kaaba, none, thought Burton, felt for the moment a deeper emotion than he. But he had to confess the humbling truth that while theirs was the high feeling of religious enthusiasm, his was but the ecstasy of gratified pride. Bare-headed and footed and in company with Mohammed, he first proceeded to the holy well, Zem-Zem, said to be the same that was shown by God to Hagar.[2] They found the water extremely unpleasant to the taste, and Burton noticed that nobody drank it without making a wry face. It was impossible at first to get near the Black Stone owing to the crush of pilgrims. However, they occupied the time in various prayers, blessed the Prophet, and kissed the finger tips of the right hand. They then made the seven Ashwat or circuits, and from time to time raised their hands to their ears, and exclaimed, "In the name of Allah, and Allah is omnipotent!" The circuits finished, it was deemed advisable to kiss the Black Stone. For

(1) Readers of *The Arabian Nights* will remember the incident in the Story of the Sweep and the Noble Lady. "A man laid hold of the covering of the Kaaba, and cried out from the bottom of his heart, saying, I beseech thee, O Allah, etc.

(2) See Genesis xxi., 15.

some minutes Burton stood looking in despair at the swarming crowd of Bedouin and other pilgrims that besieged it. But Mohammed was equal to the occasion. Noticing that most of those near the Stone were Persians, against whom the Arabs have an antipathy, he interpolated his prayers with insults directed against them—one of the mildest being "O hog and brother of a hoggess." This having small effect he collected half-a-dozen stalwart Meccans, "with whose assistance," says Burton, "by sheer strength, we wedged our way into the thin and light-legged crowd. . . . After reaching the stone, despite popular indignation testified by impatient shouts, we monopolised the use of it for at least ten minutes. While kissing it and rubbing hands and forehead upon it, I narrowly observed it, and came away persuaded that it was an aërolite." Burton and his friends next shouldered and fought their way to the part of the Kaaba called Al Multazem, at which they asked for themselves all that their souls most desired. Arrived again at the well Zem-Zem, Burton had to take another nauseous draught and was deluged with two skinfuls of the water dashed over his head. This causes sins to fall from the spirit like dust. He also said the customary prayers at the Makam Ibrahim or Praying Place of Abraham[1] and other shrines. At last, thoroughly worn out, with scorched feet and a burning head, he worked his way out of the Mosque, but he was supremely happy for he had now seen:

"Safa, Zem-Zem, Hatim's wall,
And holy Kaaba's night-black pall."[2]

The next day he journeyed to the sacred Mount of Arafat, familiar to readers of *The Arabian Nights* from the touching story of Abu Hasan and Abu Ja'afar the

(1) The stone upon which Abraham stood when he built the Kaaba. Formerly it adjoined the Kaaba. It is often alluded to in *The Arabian Nights*. The young man in The Mock Caliph says, "This is the Place and thou art Ibrahim."

(2) See also *The Arabian Nights*. The Loves of Al-Hayfa and Yusuf Burton's A.N. (Supplemental), vol. v.; Lib. Ed., vol. xi., p. 289.

Leper and[1] he estimated that he was but one of 50,000 pilgrims. The mountain was alive with people, and the huge camp at its foot had booths, huts and bazaars stocked with all manner of Eastern delicacies, and crowded with purchasers. Instead, however, of listening to the sermon, Burton got flirting with a Meccan girl with citrine skin and liquescent eyes.

On the third day, mounted on an ass, he made for Muna and took part in the ceremony called Stoning the Devil. He was, however, but one of a multitude, and, in order to get to the stoned pillar a good deal of shouldering and fighting was necessary. Both Burton and the boy Mohammed, however, gained their end, and like the rest of the people, vigorously pelted the devil, saying as they did so, "In the name of Allah—Allah is Almighty." To get out of the crowd was as difficult as it had been to get in. Mohammed received a blow in the face which brought the blood from his nose, and Burton was knocked down; but by "the judicious use of the knife" he gradually worked his way into the open again, and piously went once more to have his head shaved and his nails cut, repeating prayers incessantly. Soon after his return to Mecca, Mohammed ran up to him in intense excitement. "Rise, Effendi," he cried, "dress and follow me; the Kaaba is open." The pair then made their way thither with alacrity, and, replies to the officials in charge being satisfactory, Mohammed was authoritatively ordered to conduct Burton round the building. They entered. It was a perilous moment; and when Burton looked at the windowless walls and at the officials at the door, and thought of the serried mass of excited fanatics outside, he felt like a trapped rat. However safe a Christian might have been at Mecca, nothing could have preserved him from the ready knives of the faithful if detected in the Kaaba. The very idea was pollution to a Moslem. "Nothing," says Burton, "is more simple than the interior of this sacred building. The pavement is composed of

(1) Burton's A.N., v., 294; Lib. Ed., iv., 242.

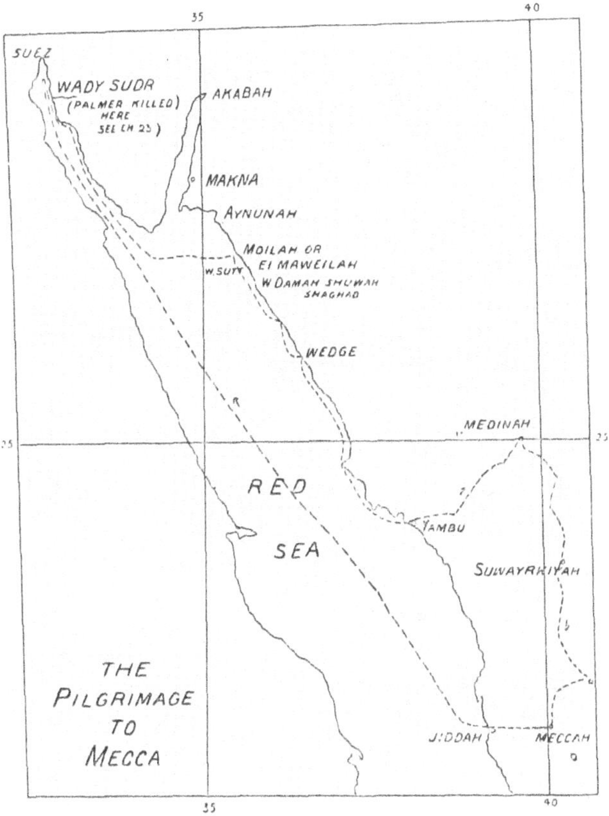
35
40
SUEZ
WADY SUDR
(PALMER KILLED)
HERE
SEE CH 23)
AKABAH
MAKNA
AYNUNAH
MOILAH OR
El MAWEILAH
W.SUTT
W DAMAH SHUWAH
SHAGHAD
WEDGE
MEDINAH
25
RED
SEA
YAMBU
SUWAYRKIYAH
JIDDAH
MECCAH
THE
PILGRIMAGE
TO
MECCA

slabs of fine and various coloured marbles. The upper part of the walls, together with the ceiling, are covered with handsome red damask, flowered over with gold. The flat roof is upheld by three cross beams, supported in the centre by three columns. Between the columns ran bars of metal supporting many lamps said to be of gold.'' The total expense was eight dollars, and when they got away, the boy Mohammed said, ''Wallah, Effendi! thou hast escaped well! some men have left their skins behind.''

The fifty-five other wonders of the city having been visited, Burton sent on Nur with his heavy boxes to Jeddah, the port of Mecca, and he himself followed soon after with Mohammed. At Jeddah he saw its one sight, the tomb of Eve, and then bade adieu to Mohammed, who returned to Mecca. Having boarded the "Dwarka,'' an English ship, he descended to his cabin and after a while emerged with all his colouring washed off and in the dress of an English gentleman. Mirza Abdullah of Bushire, ''Father of Moustaches,'' was once more Richard Francis Burton. This extraordinary exploit made Burton's name a household word throughout the world, and turned it into a synonym for daring; while his book, the *Pilgrimage to Al-Madinah and Meccah*, which appeared the following year, was read everywhere with wonder and delight. Had he been worldly-wise he would have proceeded straight to England, where, the lion of the hour, he might have obtained a reward more substantial than mere praise. But he did not show himself until the commotion caused by his exploit had been half-forgotten, and we shall find him making a similar mistake some years later, after his return from Tanganyika.[1]

It seems that Burton was known in the army as ''Ruffian Dick''—not by way of disparagement, but because of his demonic ferocity as a fighter, and because he had ''fought in single combat more enemies than

(1) See Chapter ix.

perhaps any other man of his time." One evening soon after his return from Mecca, a party of officers, including a friend of Burton's named Hawkins, were lounging outside Shepherd's Hotel at Cairo. As they sat talking and smoking, there passed repeatedly in front of them, an Arab, in his loose flowing robes, with head proudly erect, and the peculiar swinging stride of those sons of the desert. As he strode backwards and forwards he drew nearer and nearer to the little knot of officers, till at last, as he swept by, the flying folds of his burnous brushed against one of the officers. "D—— that nigger's impudence!" said the officer; "if he does that again, I'll kick him." To his surprise the dignified Arab suddenly halted, wheeled round, and exclaimed, "Well, d—— it, Hawkins, that's a fine way to welcome a fellow after two year's absence." "It's Ruffian Dick!" cried the astonished officer.[1]

Perhaps to this period must be assigned the bastinado incident. Burton used to tell the tale[2] as follows: "Once, in Egypt, another man and I were out duck shooting, and we got separated. When I next came in sight of the other man some Turkish soldiers had tied him up and were preparing to administer the bastinado. As I hurried to his assistance he said something to the Turks which I could not catch, and pointed to me. Instantly they untied him and pouncing upon me, tried to put me in his place, while my companion took to his heels. As they were six to one, they succeeded, and I had the very unpleasant experience of being bastinadoed. The first dozen or two strokes I didn't mind much, but at about the ninetieth the pain was too excruciating for description. When they had finished with me I naturally enquired what it was all for. It seems that my companion when firing at a duck had accidentally shot an Egyptian woman, the wife of one of the soldiers. Upon my appearance he had called out in Turkish to the soldiers: 'It was not I who fired the shot, it was that other fellow,' pointing

(1) *Sporting Truth.*

(2) The reader may believe as much of this story as he likes.

to me. The blackguard has taken good care to keep out of my way ever since."

The story of Burton's adventures having spread abroad, people now took the trouble to invent many incidents that were untrue. They circulated, for example, a grisly tale of a murder which he was understood to have committed on a man who had penetrated his disguise[1], and, the tale continuing to roll, the murder became eventually two murders. Unfortunately, Burton was cursed with a very foolish habit, and one that later did him considerable harm. Like Lord Byron, he delighted to shock. His sister had often reproved him for it after his return from India, but without effecting a change. Kindly listeners hardly knew how to take him, while the malicious made mischief. One day, in England, when, in the presence of his sister and a lady friend, he had thought fit to enlarge on a number of purely fictitious misdeeds, he was put to some shame. His sister having in vain tried by signs to stop him, the friend at last cut him short with: "Am I to admire you, Mr. Burton?" And he accepted the reproof. Still, he never broke himself of this dangerous habit; indeed, when the murder report spread abroad he seems to have been rather gratified than not; and he certainly took no trouble to refute the calumny.

27. Burton's Delight in Shocking.

On another occasion he boasted of his supposed descent from Louis XIV. "I should have thought," exclaimed a listener, "that you who have such good Irish blood in your veins would be glad to forget your descent from a dishonourable union."

"Oh, no," replied Burton vehemently, "I would rather be the bastard of a king than the son of an honest man."

Though this was at the time simply intended to shock, nevertheless it illustrated in a sense his real views. He used to insist that the offspring of illicit or unholy unions

(1) The man was said to have been killed in cold blood simply to silence a wagging tongue.

were in no way to be pitied if they inherited, as is often the case, the culture or splendid physique of the father and the comeliness of the mother; and instanced King Solomon, Falconbridge, in whose "large composition" could be read tokens of King Richard,(1) and the list of notables from Homer to "Pedro's son," as catalogued by Camoens(2) who said:

> "The meed of valour Bastards aye have claimed
> By arts or arms, or haply both conjoined."

The real persons to be pitied, he said, were the mentally or physically weak, whatever their parentage.

Burton now commenced to write a work to be called *El Islam*, or the History of Mohammedanism; which, however, he never finished. It opens with

28. *El Islam*.

an account of the rise of Christianity, his attitude to which resembled that of Renan.(3) Of Christ he says: "He had given an impetus to the progress of mankind by systematizing a religion of the highest moral loveliness, showing what an imperfect race can and may become." He then dilates on St. Paul, who with a daring hand "rent asunder the ties connecting Christianity with Judaism." "He offered to the great family of man a Church with a Diety at its head and a religion peculiarly of principles. He left the moral code of Christianity untouched in its loveliness. After the death of St. Paul," continues Burton, "Christianity sank into a species of idolatry. The acme of stupidity was attained by the Stylites, who conceived that mankind had no nobler end than to live and die upon the capital of a column. When things were at their worst Mohammed first appeared upon the stage of life." The work was published in its unfinished state after Burton's death.

With *The Kasîdah* we shall deal in a later chapter,

(1) See Shakespeare's *King John*, act i., scene i.

(2) Burton's translation of the *Lusiads*, vol. ii., p. 425.

(3) Although Burton began *El Islam* about 1853, he worked at it years after. Portions of it certainly remind one of Renan's *Life of Jesus*, which appeared in 1863.

CAPTAIN RICHARD BURTON.

Photo taken about 1855.

for though Burton wrote a few couplets at this time, the poem did not take its present shape till after the appearance of FitzGerald's adaptation of *The Rubaiyat Omar Khayyam.*

Having spent a few weeks in Egypt, Burton returned to Bombay, travelling in his Arab dress. Among those on board was an English gentleman, Mr. James Grant Lumsden, senior member of the Council, Bombay, who being struck by Burton's appearance, said to a friend, "What a clever, intellectual face that Arab has!" Burton, overhearing the remark, made some humorous comment in English, and thus commenced a pleasant friendship.

CHAPTER VII

29TH OCTOBER 1854—9TH FEBRUARY 1855

TO HARAR

BIBLIOGRAPHY:

13. Pilgrimage to Al-Madinah and Meccah. 3 vols. 1855–1856.

29. At Aden. *The Arabian Nights.* Oct. 1854.

It was while staying at Bombay as Mr. Lumsden's guest that Burton, already cloyed with civilization, conceived the idea of journeying, via Zaila in Somaliland, to the forbidden and therefore almost unknown city of Harar, and thence to Zanzibar. His application to the Bombay Government for permission and assistance having been received favourably, he at once set out for Aden, where he stayed with his "old and dear friend," Dr. John Steinhäuser, who had been appointed civil surgeon there. Steinhäuser, a stolid man, whose face might have been carved out of wood, was, like Burton, an enthusiastic student of *The Arabian Nights*. and their conversation naturally drifted into this subject. Both came to the conclusion that while the name of this wondrous repertory of Moslem folk-lore was familiar to almost every English child, no general reader could form any idea of its treasures. Moreover, that the door would not open to any but Arabists. But even at the present day, and notwithstanding the editions of Payne and Burton, there are still persons who imagine that *The Arabian Nights* is simply a book for the nursery. Familiar only with some inferior rendering, they are absolutely ignorant of the wealth of wisdom, humour, pathos

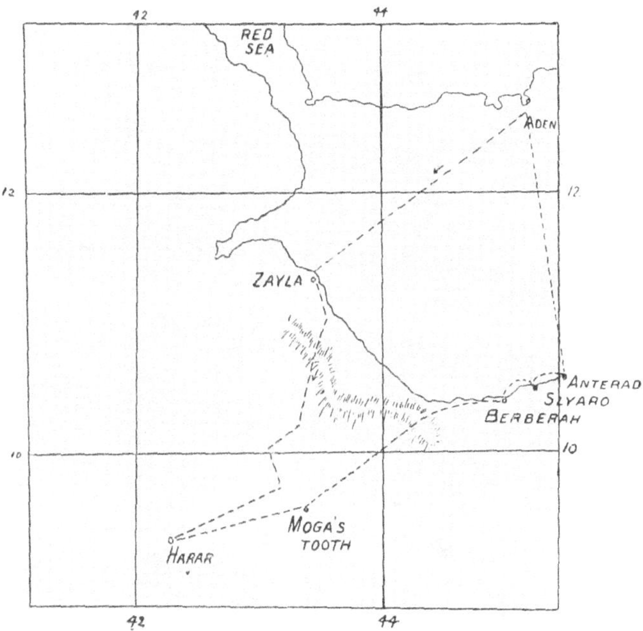

THE JOURNEY TO HARAR.

and poetry to be found in its pages.[1] Writing in 1856, Burton says: "The most familiar book in England, next to the Bible, it is one of the least known, the reason being that about one-fifth is utterly unfit for translation, and the most sanguine Orientalist would not dare to render more than three-quarters of the remainder,[2] consequently the reader loses the contrast—the very essence of the book—between its brilliancy and dulness, its moral putrefaction and such pearls as:

> 'Cast the seed of good works on the least fit soil;
> Good is never wasted, however it may be laid out.'

And in a page or two after such divine sentiment, the ladies of Baghdad sit in the porter's lap, and indulge in a facetiousness which would have killed Pietro Aretino before his time.'"[3] When the work entitled *A Thousand Nights and a Night* was commenced, no man knows. There were Eastern collections with that title four centuries ago, laboured by the bronzed fingers of Arab scribes; but the framework and some of the tales must have existed prior even to the Moslem conquest. It has been noticed that there are resemblances between the story of Shahryar and that of Ahasuerus as recorded in *Esther*. In both narratives the King is offended with his Queen and chooses a new wife daily. Shahryar has recourse to the scimitar, Ahasuerus consigns wife after wife to the seclusion of his harem. Shahryar finds a model consort in Shahrazad, Ahasuerus in Esther. Each queen saves a multitude from death, each king lies awake half the night listening to stories.[4] While many of the stories in *The Arabian Nights* are ancient, some, as internal evidence proves, are comparatively recent. Thus those of Kamar-al-Zaman II. and

(1) To some of the beauties of *The Arabian Nights* we shall draw attention in Chapter 27.

(2) Of course both Payne and Burton subsequently translated the whole.

(3) *First Footsteps in East Africa.* (The Harar Book.) Memorial Ed., p. 26.

(4) Esther, vi., 1.

Ma'aruf the Cobbler belong to the 16th century; and no manuscript appears to be older than 1548. The most important editions are the Calcutta, the Boulac [1] and the Breslau, all of which differ both in text and the order of the stories. The *Nights* were first introduced into Europe by Antoine Galland, whose French translation appeared between 1704 and 1717. Of the *Nights* proper, Galland presented the public with about a quarter, and he added ten tales [2] from other Eastern manuscripts. An anonymous English edition appeared within a few years. The edition published in 1811 by Jonathan Scott is Galland with omissions and additions, the new tales being from the Wortley Montague MS. now in the Bodleian. In 1838, Henry Torrens began a translation direct from the Arabic, of which, however, he completed only one volume, and in 1838-40 appeared the translation of Edward William Lane, [3] made direct from the Boulac edition. This work which, contains about one third of the entire *Arabian Nights*, was a great step forward, but unfortunately, Lane, who afterwards became an excellent Arabic scholar, was but a poor writer, and having no gift of verse, he rendered the poetical portions, that is to say, some ten thousand lines "in the baldest and most prosaic of English." [4]

So Burton and Steinhäuser said to themselves, As the public have never had more than one-third of the *Nights*, and that translated indifferently, we will see what we can do. "We agreed," says Burton, "to collaborate and produce a full, complete, unvarnished, uncastrated, copy of the great original, my friend taking the prose and I the metrical part; and we corresponded upon the

(1) Boulac is the port of Cairo. See Chapter xi..

(2) Zeyn al Asnam, Codadad, Aladdin, Baba Abdalla, Sidi Nouman, Cogia Hassan Alhabbal, Ali-Baba, Ali Cogia Prince Ahmed and the Fairy Peri-Banou, The two Sisters who were jealous of their Cadette.

(3) Edward William Lane (1801—1876). He is also remembered on account of his Arabic Lexicon. Five volumes appeared in 1863-74, the remainder by his grand-nephew Stanley Lane-Poole, in 1876-1890.

(4) Every student, however, must be grateful to Lane for his voluminous and valuable notes.

subject for years.[1] They told each other that, having completed their task, they would look out for a retreat as a preparation for senility, some country cottage, perhaps, in the South of France, where, remote from books, papers, pens, ink and telegrams, they could spend their nights in bed and their days in hammocks. Beyond planning the translation, and corresponding on the subject, however, nothing was done. Steinhäuser died fourteen years later (1866), and whatever notes he made were dispersed, while Burton, even as late as 1883, had done nothing beyond making a syllabus of the Boulac edition.[2] Still, the scheme was never for very long absent from his thoughts, and during his wanderings in Somaliland, the Tanganyika country and elsewhere, he often delighted the natives by reciting or reading some of the tales. The history of Burton's translation of *The Arabian Nights* is, as we shall subsequently show, curiously analogous to that of *The Kasîdah.*

30. From Zeila to Harar, 27th November 1854 to 2nd January 1855.

Burton now found that, as regards the projected expedition, his plans would have to be modified, and he finally decided to confine his explorations to "the great parched horn" of Somaliland. His plan was now to visit Harar via Zeila, and then make for Berbera, in order to join Lieutenants Speke, Herne and Stroyan, who had been authorised to assist him and had arranged to await him there. The presence at Berbera of Speke and his companions, would, it was supposed, "produce a friendly feeling on the part of Somali," and facilitate Burton's egress from Harar, should he ever, as was by no means certain, enter alive that dangerous and avoided city. Sir James Outram, then Political Resident at Aden, called the expedition a tempting of Providence, and tried hard to stop it, but in vain. Burton left Aden

(1) Lady Burton states incorrectly that the compact was made in the "winter of 1852," but Burton was then in Europe.

(2) My authorities are Mr. John Payne, Mr. Watts-Dunton and Burton's letters. See Chapter 22, § 104, and Chapter 23, § 107.

for Zeila on October 29th, taking with him a managing man called "The Hammal," a long, lean Aden policeman, nicknamed "Long Gulad" and a suave but rascally Moslem priest dubbed "The End of Time."[1] They landed on October 31st, and found Zeila a town of white-washed houses and minaretted mosques, surrounded by a low brown wall with round towers. Burton, who called himself a Moslem merchant, spent three weeks buying camels and mules and interviewing guides, while he kept up his reputation for piety with the customary devotions. According to his wont, he carefully studied the customs of the people. "One of the peculiar charms," he says, of the Somali girls, is "a soft, low and plaintive voice," and he notices that "in muscular strength and endurance the women of the Somal are far superior to their lords." The country teems with poets, who praise the persons of the belles very much in the style of Canticles, declaring prettily, for example, that their legs are as straight as the "Libi Tree," and that their hips swell out "like boiled rice." The marriage ceremonies, he tells us, are conducted with feasting, music and flogging. On first entering the nuptial hut the bridegroom draws forth his horsewhip and inflicts chastisement upon his bride, with the view of taming any lurking propensity to shrewishness. As it is no uncommon event to take four wives at once, this horsewhipping is naturally rather exhausting for the husband. Burton considered polygamy to be indispensable in countries like Somaliland, "where children are the principal wealth;" but he saw less necessity for it "among highly civilised races where the sexes are nearly equal, and where reproduction becomes a minor duty." However, he would have been glad to see polygamy allowed even in England, "if only to get rid of all the old maids," a class that he regarded with unbounded pity. He longed "to see these poor, cankered, angular ladies transformed

(1) It was prophesied that at the end of time the Moslem priesthood would be terribly corrupt.

into cheerful, amiable wives with something really to live for." "Man," it was a favourite saying with him, "is by nature polygamic, whereas woman, as a rule, is monogamic, and polyandrous only when tired of her lover. The man loves the woman, but the love of the woman is for the love of the man." He also agreed with the 18th century Rev. Martin Madan, author of *Thelyphthora*, a treatise on female ruin, who insisted that polygamy would go far to remove one of the great reproaches of the streets of London and other large cities. "Except in books," says Burton, "seduction in Mohammedan countries is almost unknown, adultery difficult." That polygamy, however, is no panacea, the following remarks will show. "Both sexes," he says, speaking of the Somali, "are temperate from necessity." Drunkenness is unknown. Still, the place is not Arcady. "After much wandering," he continues, "we are almost tempted to believe that morality is a matter of geography; [1] that nations and races have, like individuals, a pet vice; and that by restraining one, you only exasperate another. As a general rule Somali women prefer flirtations with strangers, following the well-known Arabian proverb, 'The new comer filleth the eye.'" Burton was thoroughly at home in Zeila "with the melodious chant of the muezzin" and the loudly intoned "Amin" and "Allaho Akbar" daily ringing in his ear. He often went into the Mosque, and with a sword and a rosary before him, read the "cow chapter" [2] in a loud twanging voice. Indeed, he had played the rôle of devout Mohammedan so long, that he had almost become one. The people of Zeila tried to persuade him to abandon his project. "If," said they, "you escape the desert hordes it will only be to fall by the hands of the truculent Amir of Harar." Nothing, however, could dash Burton's confidence in his star, and

(1) Later he was thoroughly convinced of the soundness of this theory. See Chapters xxii. to xxx.

(2) In the Koran.

like Dante, he applied to Fear no epithets but "vile" and "base."

One Raghi, a petty Eesa chief, having been procured as protector of the party, and other arrangements having been made, Burton on November 27th (1854) set out for his destination by a circuitous route. Raghi rode in front. Next, leading camels, walked two enormously fat Somali women; while by the side of the camels rode Burton's three attendants, the Hammal, Long Gulad, and "The End of Time," "their frizzled wigs radiant with grease," and their robes splendidly white with borders dazzlingly red. Burton brought up the rear on a fine white mule with a gold fringed Arab pad and wrapper-cloth, a double-barrelled gun across his lap, and in this manner the little caravan pursued its sinuous course over the desert. At halting places he told his company tales from *The Arabian Nights*; they laughed immoderately at the adventures of the little Hunchback; tears filled their eyes as they listened to the sad fate of Azizah;[1] and the two fat Somali women were promptly dubbed Shahrazad and Dunyazad. Dunyazad had been as far as Aden and was coquettish. Her little black eyes never met Burton's, and frequently with affected confusion she turned her sable cheek the clean contrary way. Attendant on the women was a Zeila lad, who, being one-eyed, was pitilessly called "The Kalandar." At their first halting place, Burton astonished the natives by shooting a vulture on the wing. "Lo!" cried the women, "he bringeth down the birds from heaven." On their way through an ochreish Goban, or maritime plain, they passed huge hills made by white ants, Gallas graves planted with aloe,[2] and saw in the distance troops of gazelles. They were now in the Isa country, "Traitorous as an Isa" being a Zeila proverb. Though the

(1) Burton's A.N., ii. 323; Lib. Ed., ii., p. 215.

(2) When the aloe sprouts the spirits of the deceased are supposed to be admitted to the gardens of Wak (Paradise). *Arabian Nights*, Lib. Ed., i., 127.

people were robbers and murderers, Burton, by tact, got on excellently with them, and they good-naturedly offered him wives. At every settlement the whole population flocked to see him, the female portion loudly expressing their admiration for him. "Come girls," they cried one to another, "come and look at this white stranger." According to Raghi, the fair face of a French lady who had recently landed at Berbera, "made every man hate his wife, and every wife hate herself." Once they were attacked by Bedouin, who, however, on hearing the report of Burton's revolver, declared that they were only in fun. Others who tried to stop them were shown the star sapphire, and threatened with "sorcery, death, wild beasts," and other unpleasantnesses. At a place called Aububah, Raghi relinquished the charge of the caravan to some men of the Gudabirsi tribe, who led the way to the village of Wilensi, where they were the guests of the household of a powerful chief called Jirad Adan. Here Burton left Shahrazad, Dunyazad and the Kalandar, and proceeded to Sagharrah, where he met and formed a friendship with Jirad Adan. For several days he was prostrated by fever, and some Harar men who looked in tried to obtain him as a prisoner. The Jirad acted honourably, but he declined to escort Burton to Harar. "No one," he said, "is safe in the Amir's clutches, and I would as soon walk into a crocodile's mouth as set foot in the city." "Nothing then remained," says Burton, "but *payer d'audace*,[1] and, throwing all forethought to the dogs, to rely upon what has made many a small man great, the good star. I addressed my companions in a set speech, advising a mount without delay."[2] The End of Time, having shown the white feather, was left behind, but the rest courageously consented to accompany their leader. "At 10 a.m. on the 2nd January," says Burton, "all the villagers assembled, and recited the

(1) To face it out.

(2) *First Footsteps in East Africa*, i. 196.

Fatihah, consoling us with the information that we were dead men.'' The little company, carrying their lives in their hands, then set forward, and presently came in sight of Harar, "a dark speck upon a tawny sheet of stubble.'' Arrived at the gate of the town, they accosted the warder, sent their salaams to the Amir, and requested the honour of audience.

31. At Harar.

They were conducted to the palace, a long, single-storied, windowless barn of rough stone and reddish clay. Says Burton: "I walked into a vast hall between two long rows of Galla spearmen, between whose lines I had to pass. They were large, half-naked savages, standing like statues with fierce, movable eyes, each one holding, with its butt end on the ground, a huge spear, with a head the size of a shovel. I purposely sauntered down them coolly with a swagger, with my eyes fixed upon their dangerous-looking faces. I had a six-shooter concealed in my waist-belt, and determined, at the first show of excitement, to run up to the Amir, and put it to his head, if it were necessary, to save my own life.'' The Amir was an etiolated young man of twenty-four or twenty-five, plain and thin-bearded, with a yellow complexion, wrinkled brows and protruding eyes. He wore a flowing robe of crimson cloth, edged with snowy fur, and a narrow white turban tightly twisted round a tall, conical cap of red velvet. On being asked his errand, Burton replied politely in Arabic that he had come from Aden in order to bear the compliments of the governor, and to see the light of his highness's countenance. On the whole, the Amir was gracious, but for some days Burton and his party were in jeopardy, and when he reflected that he was under the roof of a bigoted and sanguinary prince, whose filthy dungeons resounded with the moans of heavily ironed, half-starved prisoners; among a people who detested foreigners; he, the only European who had ever passed over their inhospitable threshold, naturally felt uncomfortable. The Amir, it seems, had four principal wives, and an army of 200 men armed chiefly with daggers. Burton describes the streets of Harar as dirty narrow lanes heaped with garbage, and the houses as situated at the bottom of courtyards, closed by gates of holcus stalks. The town was proud of its learning and sanctity, and venerated the memory of several very holy and verminous saints. Neither sex possessed personal attractions, and the head-dresses of the women seen from behind resembled a pawnbroker's sign, except that they were blue instead of gilt. The people lived chiefly on holcus, and a narcotic called "jat," made by pounding the tender twigs of a tree of the same name. "It produced in them," "says Burton, "a manner of dreamy enjoyment, which exaggerated by time and distance, may have given rise to that splendid myth the Lotos and the Lotophagi.(1) Their chief commodity was coffee, their favourite drink an aphrodisiac made of honey dissolved in hot water, and strained and fermented with the bark of a tree called kudidah. Although unmolested, Burton had no wish to remain long at Harar, and when on 13th January he and his party took their departure it was with a distinct feeling of relief.

At Sagharrah they found again the pusillanimous "End of Time," and at Wilensi they were rejoined by Shahrazad, Dunyazad, and the one-eyed Kalandar. Persons who met Burton and his friends enquired Irish-like if they were the party who had been put to death by the Amir of Harar. Everyone, indeed, was amazed to see them not only alive, but uninjured, and the Frank's temerity became the talk of the desert. Burton now put the two women, the Kalandar, the camels, and the baggage, under the care of a guide, and sent them to Zeila, while he himself and the men made straight for Berbera. The journey, which led them past Moga's tooth(2) and

32. From Harar to Berbera. 13th Jan. 1855—5th Feb. 1855.

(1) *First Footsteps in East Africa*, ii., 31.

(2) The legend of Moga is similar to that of Birnam Wood's March, used by Shakespeare in *Macbeth*.

Gogaysa, was a terrible one, for the party suffered tortures from thirst, and at one time it seemed as though all must perish. By good fortune, however, they ultimately came upon some pools. Any fear that might have haunted them, lest the water should be poisonous, was soon dispelled, for it contained a vast number of tadpoles and insects, and was therefore considered quite harmless and suitable for drinking. For many hours they again plodded on beneath a brazen sky. Again thirst assailed them; and, like Ishmael in the desert of Zin, they were ready to cast themselves down and die. This time they were saved by a bird, a katta or sand grouse, which they saw making for some hills; and having followed it, they found, as they had anticipated, a spring of water, at which they frenziedly slaked their thirst. Many other difficulties and troubles confronted them in their subsequent march, but at last they heard (delightful sound!) the murmur of the distant sea. Every man was worn out, with the exception of the Hammal, who, to Burton's delight, not only talked, but sang and shouted. Finally they reached Berbera, where they found Speke, Herne and Stroyan, and on 5th February, Burton in company with the Hammal, Long Gulad, and The End of Time, set sail for Aden, calling on their way at Siyaro and Anterad, east of Berbera.

The first news Burton had on arriving there was of the death of his mother, which had occurred 18th December 1854, at the time he lay ill at Sagharrah. Always immersed in him, she used to say, when he left her, "It seems as if the sun itself has disappeared." He, on his part, often bore witness to the unselfishness and blamelessness of her life, generally adding, "It is very pleasant to be able to feel proud of one's parents."

33. The Fight at Berbera, 22nd April, 1855.

Unable to let well alone, Burton now wanted to make a new expedition, this time to the Nile, via Berbera and Harar, and on a larger and more imposing scale. On 7th April he was back again at Berbera, taking with him Speke, Stroyan, Herne and 42 assistants, and his first care was to establish an agency on the coast, so as to have the protection of the English gunboat, the "Mahi," which had brought them. Unfortunately, the Government drew off the gunboat, and this had scarcely been done before Burton and his party were attacked by 300 natives, who swarmed round them during the night, and tried to entrap and entangle them by throwing down the tents. A desperate hand-to-hand fight then ensued. Javelins hissed, war-clubs crashed, The forty-two coloured auxiliaries promptly took to their heels, leaving the four Englishmen to do as they could. Stroyan fell early in the fight. Burton, who had nothing but a sabre, fought like a demon ; Speke, on his left near the entrance of the tent, did deadly execution with a pair of revolvers ; Herne on his right emptied into the enemy a sixshooter, and then hammered it with the butt end. Burton, while sabreing his way towards the sea, was struck by a javelin, which pierced both cheeks, and struck out four of his teeth. Speke received eleven wounds, from which, however, he took no harm—a touching proof, comments Burton, of how difficult it is to kill a man in sound health. Eventually the survivors, stained with blood, and fearfully exhausted, but carrying, nevertheless, the corpse of poor Stroyan, managed to reach a friendly native craft, which straightway took them back to Aden.[1]

[1] The story of these adventures is recorded in *First Footsteps in East Africa*, dedicated to Lumsden, who, in its pages, is often apostrophised as "My dear L."

CHAPTER VIII.

9TH FEBRUARY 1855—OCTOBER 1856.

THE CRIMEA.

BIBLIOGRAPHY

14. First Footsteps in East Africa, 1856.

Owing to his wounds Burton had to return to England, and, on his first opportunity, he gave an account of his explorations before the Royal Geographical Society. Little, however, was now talked of except the Crimean War, which had commenced, it will be remembered, in March 1854. The Allies landed in the Crimea in September, Inkermann was fought on the 5th of November, and then followed the tedious siege of Sebastopol. Burton had not long been home before he applied for and obtained leave to join the besieging army; and his brother Edward also went out as surgeon, about the same time. Emulous of the deeds of Napier and Outram, Burton now thought he saw a career of military glory awaiting him. Soon after his arrival at the seat of war he was appointed chief of the staff to General Beatson, and in his "gorgeous uniform blazing with gold" he set vigorously to work to re-organise and drill his contingent of Bashi-Bazouks. He had great difficulties with Beatson, a brave, but passionate and undiplomatic old warrior; but he succeeded marvellously with his men, and his hope of winning fame rose higher than ever. The war, however, was crawling to an end, and the troops he had drilled so patiently had little to do beside look on. At this conjuncture he thought he saw a road to success in the relief

34. The Crimea.

of Kars, which had been persistently besieged by the Russians. Elated at the prospect of taking part in a great military feat, he hurried to Constantinople, obtained an interview with the British Ambassador, Lord Stratford, and submitted a plan for approval. To his amazement, Lord Stratford broke into a towering passion, and called him " the most impudent man in the Bombay Army." Later Burton understood in what way he had transgressed. As the war was closing, it had been arranged by the Allies that Kars should be allowed to fall as a peace offering to Russia.

Burton now began to suffer from the untrue tales that were told about him, still he never troubled to disprove them. Some were circulated by a fellow officer of his—an unmitigated scoundrel whose life had been sullied by every species of vice; who not only invented calumniating stories but inserted particulars that gave them a verisimilitude. Two of this man's misdeeds may be mentioned. First he robbed the Post Office at Alexandria, and later he unblushingly unfolded to Lord Stanley of Alderley his plan of marrying an heiress and of divorcing her some months later with a view to keeping, under a Greek law, a large portion of her income. He seemed so certain of being able to do it that Lord Stanley consulted a lady friend, and the two together succeeded in frustrating the infamous design. This sordid and callous rascal tried hard to lead people to suppose that he and Burton were hand and glove in various kinds of devilry, and a favourite phrase in his mouth was " I and Burton are great scamps." Percy Smythe[1] then an official under Lord Stratford, commented on hearing the saying: " No, that won't do, —— is a real scamp, but Burton is only wild." One story put abroad apparently by the same scoundrel is still in circulation. We are told that Burton was once caught in a Turkish harem, and allowed to escape only after suffering the usual

(1) Afterwards Lord Strangford. The correspondence on this subject was lent me by Mr. Mostyn Pryce, who received it from Miss Stisted.

indescribable penalty. As this was the solitary story that really annoyed Burton, we think it our duty to say that conclusive documentary evidence exists proving that, whether or not he ever broke into a harem, he most certainly underwent no deprivation. Other slanders of an even more offensive nature got abroad. Pious English mothers loathed Burton's name, and even men of the world mentioned it apologetically. In time, it is true, he lived all this down, still he was never—he is not now—generally regarded as a saint worthy of canonization.

With the suspension of General Beatson—for the machinations of enemies ultimately accomplished the old hero's fall—Burton's connection with the Crimean army abruptly ceased. Having sent in his resignation, he returned to England and arrived here just in time to miss, to his disappointment, his brother Edward, who had again left for Ceylon. Edward's after career was sad enough to draw tears from adamant. During an elephant hunt a number of natives set upon him and beat him brutally about the head. Brain trouble ensued, and he returned home, but henceforth, though he attained a green old age, he lived a life of utter silence. Except on one solitary occasion he never after—and that is to say for forty years—uttered a single word. Always resembling a Greek statue, there was now added to him the characteristic of all statues, rigid and solemn silence. From a man he had become aching marble. To Burton, with his great. warm, affectionate heart, Edward's affliction was an unceasing grief. In all his letters he enquires tenderly after his "dear brother," and could truly say, with the enemy of his boyhood, Oliver Goldsmith :

> " Where'er I roam, whatever realms to see,
> My heart untravelled fondly turns to thee :
> Still to my Brother turns." (1)

Arrived in England, General Beatson promptly instituted civil proceedings against his enemies ; and Burton

(1) *The Traveller.*

MAJOR EDWARD J. N. BURTON

(SIR RICHARD BURTON'S BROTHER.)

was in constant expectation of being subpœnaed. He thoroughly sympathized with Beatson, but he had no wish to be forced to remain in London, just as he had no wish at any time in his life to be mewed up anywhere. Consequently he disguised himself by wearing green spectacles and tying a pillow over his stomach to simulate corpulence. To one friend who met him, he made himself known. "Are you really Burton?" inquired his friend. "I shall be," replied Burton, "but just now I'm a Greek doctor." Burton's conscience, however, finally had the mastery. He did attend the trial and he corroborated the statements of his late chief. The verdict of the jury went against Beatson, but it was generally felt that the old war dog had fully vindicated his character.

35. Engaged to Isabel Arundell, August 1856.

In August, after a lapse of four years, Burton renewed acquaintance with Isabel Arundell, who one day met him, quite by accident, in the Botanical Gardens, and she kept meeting him there quite by accident every day for a fortnight. He had carried his life in his hand to Mecca and to Harar, he had kept at bay 200 Somalis, but like the man in Camoens, he finally fell by "a pair of eyes."[1] According to Lady Burton,[2] it was Burton who made the actual proposal; and it is just possible.

"You won't chalk up 'Mother will be angry' now I hope," said Burton.

"Perhaps not," replied Miss Arundell, "but she will be all the same."

Mrs. Arundell, indeed, like so many other English mothers, was violently prejudiced against Burton. When her daughter broached the subject she replied fiercely: "He is not an old English Catholic, or even a Catholic, he has neither money nor prospects." She might also have added that he was apt to respect mere men of

(1) Burton's *Camoens*, ii., 445.

(2) The marriage did not take place till 22nd January 1861. See Chapter x.

intellect more than men of wealth and rank, an un-English trait which would be sure to militate against his advancement.

Miss Arundell bravely defended her lover, but without effect. A few days later she again met her old gipsy crone Hagar Burton, who repeated her sibylline declaration. As Miss Arundell never, by any chance, talked about anything or anybody except Burton, and as she paid liberally for consulting the Fates, this declaration necessarily points to peculiar acumen on the part of the gipsy.

At one of their meetings Miss Arundell put round Burton's neck a steel chain with a medal of the Virgin Mary and begged him to wear it all his life. Possessing a very accommodating temperament in matters that seemed to himself of no vital importance, he consented: so it joined the star-sapphire and other amulets, holy and unholy, which, for different purposes, he carried about the world.

That this medal had often acted as a preservative to Burton she was in after life thoroughly convinced.

CHAPTER IX

THE UNVEILING OF ISIS

DECEMBER 1856—21ST MAY 1859

BIBLIOGRAPHY:

15. Lake Regions of Equatorial Africa.
16. Vol. 33 of the Royal Geographical Society.

The fame of a soldier having been denied him, Burton now turned his thoughts once more to exploration; and his eagerness for renown is revealed conspicuously in some verses written about this time. They commence:

36. To Fuga. January to March 1857.

> "I wore thine image, Fame,
> Within a heart well fit to be thy shrine!
> Others a thousand boons may gain;
> One wish was mine."

He hoped to obtain one of its smiles and then die. A glorious hand seemed to beckon him to Africa. There he was to go and find his destiny. The last stanza runs:

> "Mine ear will hear no other sound,
> No other thought my heart will know.
> Is this a sin? Oh, pardon, Lord!
> Thou mad'st me so."

He would obtain the fame of a great traveller; the earth should roll up for him as a carpet. Happy indeed was Isabel Arundell when he placed the verses in her hand, but melancholy to relate, he also presented copies to his "dear Louisa," and several other dears.

He now read greedily all the great geographers, ancient and modern, and all the other important books bearing on African exploration. If he became an authority on

Herodotus, Pliny, Ptolemy, Strabo, and Pomponius Mela, he became equally an authority on Bruce, Sonnini, Lacerda, the Pombeiros, Monteiro and Gamitto.

From Ptolemy downwards writers and travellers had prayed for the unveiling of Isis, that is to say, the discovery of the sources of the Nile ; but for two thousand years every effort had proved fruitless. Burning to immortalize himself by wresting from the mysterious river its immemorial secret, Burton now planned an expedition for that purpose. Thanks to the good offices of Lord Clarendon, Secretary of State for Foreign Affairs, the Royal Geographical Society promised him the necessary funds ; while Cardinal Wiseman, ever his sincere friend, gave him a passport to all Catholic missionaries [1] To Burton, as we have seen, partings were always distressing, and in order to avoid bidding adieu to Miss Arundell he adopted his usual course, leaving a letter which mentioned love and that he was gone.

He quitted England for Bombay in October 1856, and crossed to Zanzibar in the *Elphinstone* sloop of war, Speke, who was to be his companion in the expedition, sailing with him. Burton was in the highest spirits. "One of the gladdest moments in human life," he wrote, "is the departing upon a distant journey into unknown lands. Shaking off with one effort the fetters of habit, the leaden weight of routine, the slavery of civilisation,[2] man feels once more happy. The blood flows with the fast circulation of youth, excitement gives a new vigour to the muscles and a sense of sudden freedom adds an inch to the stature." Among the crew was a midshipman, C. R. Low, who became a life-long friend of Burton. Says Mr. Low, "We used to have bouts of single-stick in the pleasant evenings in the poop, and many's the time he has blacked my arms and legs with his weapons. . . . . Though a dangerous enemy, he was a warm and constant

(1) This is now in the public library at Camberwell.

(2) In England men are slaves to a grinding despotism of conventionalities. *Pilgrimage to Meccah*, ii., 86.

friend."[1] On reaching Zanzibar, Burton, finding the season an unsuitable one for the commencement of his great expedition, resolved to make what he called "a preliminary canter." So he and Speke set out on a cruise northward in a crazy old Arab "beden" with ragged sails and worm-eaten timbers. They carried with them, however, a galvanised iron life-boat, "The Louisa," named after Burton's old love, and so felt no fear.

They passed the Island of Pemba, and on the 22nd reached Mombasa, which Burton was glad to visit on account of its associations with Camoens, who wrote:

> So near that islet lay along the land,
> Nought save a narrow channel stood atween;
> And rose a city thronèd on the strand,
> Which from the margent of the seas was seen;
> Fair built with lordly buildings tall and grand,
> As from its offing showèd all its sheen,
> Here ruled a monarch for long years high famed,
> Islet and city are Mombasa named.[2]

Indeed he never missed an opportunity of seeing spots associated with his beloved "Master." Then they turned southward and on February 3rd reached Pangany, whence, in company with a facetious fellow named Sudy Bombay, they set out on a canoe and foot journey to Fuga, which they found to be "an unfenced heap of hay cock huts." Though a forbidden city to strangers they managed to get admittance by announcing themselves as "European wizards and *Waganga* of peculiar power over the moon, the stars, the wind and the rain." They found the sultan of the place, an old man named Kimwere, sick, emaciated and leprous. He required, he said, an elixir which would restore him to health, strength, and youth. This, however, despite his very respectable knowledge of medicine, Burton was not able to compound,

(1) Unpublished letter to Miss Stisted, 23rd May 1896.

(2) We have given the stanza in the form Burton first wrote it—beginning each line with a capital. The appearance of Mombasa seems to have been really imposing in the time of Camoens. Its glory has long since departed.

so after staying two days he took his leave. "It made me sad," says Burton, "to see the wistful, lingering look with which the poor old king accompanied the word Kuahery! (Farewell!)" On the return journey Speke shot a hippopotamus which he presented to the natives, who promptly ate it. By the time Pangany was again reached both travellers were in a high fever; but regarding it simply as a seasoning, they felt gratified rather than not. When the Zanzibar boat arrived Speke was well enough to walk to the shore, but Burton "had to be supported like a bedridden old woman."

37. Zanzibar to Tanganyika. 26th June 1857 to 26th May 1858.

Burton left Zanzibar on his great expedition at the end of June, carrying with him various letters of introduction from the Sultan of Zanzibar, a diploma signed by the Shaykh El Islam of Mecca, and the passport already mentioned of Cardinal Wiseman. To his star-sapphire he added some little canvas bags containing horse chestnuts which he carried about "against the Evil Eye, and as a charm to ward off sickness."(1) Beside Burton and Speke, the party consisted of two Goa boys, two negro gun-carriers, Sudy Bombay, and ten Zanzibar mercenaries. Dr. Steinhäuser, who had hoped to join them, was restrained by illness. "My desire," says Burton, "was to ascertain the limits of Tanganyika Lake, to learn the ethnography of its tribes, and to determine the export of the produce of the interior," He held the streams that fed Tanganyika to be the ultimate sources of the Nile; and believed that the glory of their discovery would be his. Fortune, however, the most fickle of goddesses, thought fit to deprive him of this ardently coveted boon.

The explorers landed at Wale Point on June 26th, and on July 14th reached K'hutu. At Dut'humi Burton, despite his bags of chestnuts, fell with marsh fever, and in his fits he imagined himself to be "two persons who were

(1) These little bags were found in his pocket after his death. See Chapter xxxviii.

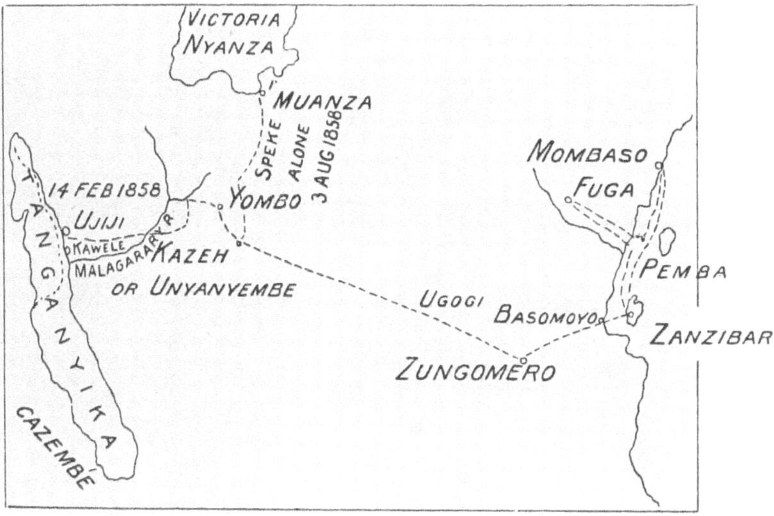

THE JOURNEY TO TANGANYIKA.

inimical to each other,'' an idea very suitable for a man nursing the "duality" theory. When he recovered, fresh misfortunes followed, and finally all the riding asses died. Burton, however, amid it all, managed to do one very humane action. He headed a little expedition against a slave raider, and had the satisfaction of restoring five poor creatures to their homes.

The tropical vegetation and the pleasant streams afforded delightful vistas both by daylight and moonlight, but every mile the travellers were saddened by the sight of clean-picked skeletons or swollen corpses. Sometimes they met companies of haggard, heavy-gaited men and women half blind with small-pox—the mothers carrying on their backs infants as loathsome as themselves. Near every kraal stood detached huts built for the diseased to die in. They passed from this God-forsaken land to a district of springs welling with sweet water, calabashes and tamarinds, and circlets of deep, dew-fed verdure. The air was spicy, and zebras and antelopes browsed in the distance. Then the scene again changed, and they were in a slimy, malarious swamp. They were bitten by pismires an inch long, and by the unmerciful tzetze fly. The mercenaries, who threatened to desert, rendered no assistance, and the leader, one Said bin Salim, actually refused to give Burton a piece of canvas to make a tent. Sudy Bombay then made a memorable speech, "O Said," he said, "if you are not ashamed of your master, be at least ashamed of his servant," a rebuke that had the effect of causing the man to surrender at once the whole awning. At other times the star-sapphire which Burton carried on his person proved a valuable auxiliary—and convinced where words failed. But the mercenaries, mistaking Burton's forbearance for weakness, became daily bolder and more insolent, and they now only awaited a convenient opportunity to kill him. One day as he was marching along, gun over shoulder and dagger in hand, he became conscious that two of his men were unpleasantly near, and after a while one of them, unaware that Burton understood his language, urged the other to strike. Burton did not hesitate a moment. Without looking round, he thrust back his dagger, and stabbed the man dead on the spot.[1] The other, who fell on his knees and prayed for mercy, was spared. This, however, did not cure his followers of their murderous instincts, and a little later he discovered another plot. The prospective assassins having piled a little wood where they intended to kindle a fire, went off to search for more. While they were gone Burton made a hole under the wood and buried a canister of gunpowder in it. On their return the assassins lighted the fire, seated themselves comfortably round, and presently there weren't any assassins. We tell these tales just as Burton told them to his intimate friends. The first may have been true, the second, we believe, simply illustrates his inveterate habit of telling tales against himself with the desire to shock. In any circumstances, his life was in constant peril; but he and the majority of the party, after unexampled tortures from thirst, arrived footsore and jaded in a veritable land of Goshen—Kazeh or Unyanyembe, where they met some kindly Arab merchants.

"What a contrast," exclaims Burton, "between the open-handed hospitality and the hearty good-will of this noble race—the Arabs—and the niggardliness of the savage and selfish African. It was heart of flesh after heart of stone." Burton found the Arabs of Kazeh living comfortably and even sybaritically. They had large, substantial houses, fine gardens, luxuries from the coast and "troops of concubines and slaves." Burton gallantly gives the ladies their due. "Among the fair of Yombo," he says, "there were no fewer than three beauties—women who would be deemed beautiful in any part of the world. Their faces were purely Grecian; they had laughing eyes, their figures were models for an artist with—

"Turgide, brune, e ritondette mamme,"

(1) This story nowhere appears in Burton's books. I had it from Mr. W. F. Kirby, to whom Burton told it.

like the 'bending statue' that delights the world. The dress—a short kilt of calabash fibre—rather set off than concealed their charms, and though destitute of petticoat they were wholly unconscious of indecorum. These beautiful domestic animals graciously smiled when in my best Kenyamwezi I did my devoir to the sex; and the present of a little tobacco always secured for me a seat in the undress circle."

Of the native races of West Africa Burton gave a graphic account when he came to write the history of this expedition.[1] All, it seems, had certain customs in common. Every man drank heavily, ate to repletion and gambled. They would hazard first their property and then themselves. A negro would stake his aged mother against a cow. As for morality, neither the word nor the thing existed among them. Their idea of perfect bliss was total intoxication. When ill, they applied to a medicine man, who having received a fee used it for the purpose of getting drunk, but upon his return to sobriety, he always, unless, of course, the patient took upon himself to die instead of waiting, attended conscientiously to his duties. No self-respecting chief was ever sober after mid-day. Women were fattened for marriage just as pigs are fattened for market—beauty and obesity being interchangeable terms. The wearisome proceedings in England necessary to a divorce, observes Burton, are there unknown. You turn your wife out of doors, and the thing is done.

The chief trouble at Kazeh, as elsewhere, arose from the green scorpion, but there were also lizards and gargantuan spiders. Vermin under an inch in length, such as fleas, ants, and mosquitoes, were deemed unworthy of notice. The march soon began again, but they had not proceeded many miles before Burton fell with partial paralysis brought on by malaria; and Speke, whom Burton always called "Jack," became partially blind. Thoughts of the elmy fields and the bistre furrows of

[1] *The Lake Regions of Central Africa*, 1860.

Elstree and the tasselled coppices of Tours crowded Burton's brain; and he wrote:

> "I hear the sounds I used to hear,
> The laugh of joy, the groan of pain,
> The sounds of childhood sound again
> Death must be near."

At last, on the 13th February they saw before them a long streak of light. "Look, master, look," cried Burton's Arab guide, "behold the great water!" They advanced a few yards, and then an enormous expanse of blue burst into sight. There, in the lap of its steel-coloured mountains, basking in the gorgeous tropical sunshine, lay the great lake Tanganyika. The goal had been reached; by his daring, shrewdness and resolution he had overcome all difficulties. Like the soldiers in Tacitus, in victory he found all things—health, vigour, abundance.

No wonder Burton felt a marvellous exultation of spirits when he viewed this great expanse of waters. Here, he thought, are the sources of that ancient river—the Nile. Now are fulfilled the longings of two thousand years. I am the heir of the ages! Having hired "a solid built Arab craft," the explorers made their way first to Ujiji and then to Uvira, the northernmost point of the lake, which they reached on April 26th. On their return voyage they were caught in a terrible storm, from which they did not expect to be saved, and while the wild tumbling waves threatened momentarily to engulf them a couplet from his fragmentary *Kasîdah* kept running in Burton's mind:

> "This collied night, these horrid waves, these gusts that sweep the whirling deep;
> What reck they of our evil plight, who on the shore securely sleep?" (1)

However, they came out of this peril, just as they had

(1) Subsequently altered to "This gloomy night, these grisly waves, etc." The stanza is really borrowed from Hafiz. See Payne's Hafiz, vol. i., p. 2.

> Dark the night and fears possess us, Of the waves and whirlpools wild:
> Of our case what know the lightly Laden on the shores that dwell?"

come out of so many others. Burton also crossed the lake and landed in Kazembé's country,[1] in which he was intensely interested, and some years later he translated into English the narratives of Dr. Lacerda[2] and other Portuguese travellers who had visited its capital, Lunda, near Lake Moero.

38. The Return Journey, 26th May 1858 to 13th February 1859.

The explorers left Tanganyika for the return journey to Zanzibar on May 26th. At Yombo, reached June 18th, Burton received a packet of letters, which arrived from the coast, and from one he learnt of the death of his father, which had occurred 8 months previous. Despite his researches, Colonel Burton was not missed in the scientific world, but his son sincerely mourned a kind-hearted and indulgent parent. At Kazeh, Fortune, which had hitherto been so favourable, now played Burton a paltry trick. Speke having expressed a wish to visit the lake now called Victoria Nyanza, a sheet of water which report declared to be larger than Tanganyika, Burton, for various reasons, thought it wiser not to accompany him. So Speke went alone and continued his march until he reached the lake, the dimensions of which surpassed his most sanguine expectations. On his return to Kazeh he at once declared that the Victoria Nyanza and its affluents were the head waters of the Nile, and that consequently he had discovered them. Isis (he assured Burton) was at last unveiled. As a matter of fact he had no firmer ground for making that statement than Burton had in giving the honour to Tanganyika, and each clung tenaciously to his own theory. Speke, indeed, had a very artistic eye. He not only, by guess, connected his lake with the Nile, but placed on his map a very fine range of mountains which had no existence—the Mountains of the Moon. However, the fact remains that as regards the Nile his theory turned out to be the correct one. The expedition

(1) The ruler, like the country, is called Kazembé.

(2) Dr. Lacerda died at Lunda 18th October 1798. Burton's translation, *The Lands of the Cazembe*, etc., appeared in 1873.

went forward again, but his attitude towards Burton henceforth changed. Hitherto they had been the best of friends, and it was always "Dick" and "Jack," but now Speke became querulous, and the mere mention of the Nile gave him offence. Struck down with the disease called "Little Irons," he thought he was being torn limb from limb by devils, giants, and lion-headed demons, and he made both in his delirium and after his recovery all kinds of wild charges against Burton, and interlarded his speech with contumelious taunts—his chief grievance being Burton's refusal to accept the Victoria Nyanza-Nile theory. But Burton made no retort. On the contrary, he bore Speke's petulance with infinite patience. Perhaps he remembered the couplet in his favourite *Beharistan:*

> "True friend is he who bears with all
> His friend's unkindness, spite and gall."(1)

There is no need for us to side either with Speke or Burton. Both were splendid men, and their country is proud of them. Fevers, hardships, toils, disappointments, ambition, explain everything, and it is quite certain that each of the explorers inwardly recognised the merit of the other. They reached Zanzibar again 4th March 1859.

Had Burton been worldly wise he would have at once returned home, but he repeated the mistake made after the journey to Mecca and was again to suffer from it.

Speke, on the other hand, who ever had an eye to the main chance, sailed straight for England, where he arrived 9th May 1859. He at once took a very unfair advantage of Burton "by calling at the Royal Geographical Society and endeavouring to inaugurate a new exploration" without his old chief. He was convinced, he said, that the Victoria Nyanza was the source of the Nile, and he wished to set the matter at rest once and for ever by visiting its northern shores. The Society joined with him Captain James A. Grant (2) and it was settled that this new

(1) *The Beharistan.* 1st Garden.

(2) J. A. Grant, born 1827, died 10th February, 1892.

expedition should immediately be made. Speke also lectured vaingloriously at Burlington House. When Burton arrived in London on May 21st it was only to find all the ground cut from under him. While Speke, the subordinate, had been welcomed like a king, he, Burton, the chief of the expedition, had landed unnoticed. But the bitterest pill was the news that Speke had been appointed to lead the new expedition. And as if that was not enough, Captain Rigby, Consul at Zanzibar, gave ear to and published the complaints of some of Burton's dastardly native followers. Although Fortune cheated Burton of having been the actual discoverer of the Source of the Nile, it must never be forgotten that all the credit of having inaugurated the expedition to Central Africa and of leading it are his. Tanganyika—in the words of a recent writer, "is in a very true sense the heart of Africa." If some day a powerful state spring up on its shores, Burton will to all time be honoured as its indomitable Columbus. In his journal he wrote proudly, but not untruly: "I have built me a monument stronger than brass." The territory is now German. Its future masters who shall name! but whoever they may be, no difference can be made to Burton's glory. Kingdoms may come and kingdoms may go, but the fame of the truly great man speeds on for ever

CHAPTER X.

22ND JANUARY 1861—TO AUGUST 1861

MORMONS AND MARRIAGE

BIBLIOGRAPHY:

17. The City of the Saints, 1861.

39. We rushed into each other's arms. 22nd May, 1860.

During Burton's absence Isabel Arundell tortured herself with apprehensions and fears. Now and again a message from him reached her, but there were huge deserts of silence. Then came the news of Speke's return and lionization in London. She thus tells the story of her re-union with Burton. "On May 22nd (1860), I chanced to call upon a friend. I was told she had gone out, but would be in to tea, and was asked to wait. In a few minutes another ring came to the door, and another visitor was also asked to wait. A voice that thrilled me through and through came up the stairs, saying, 'I want Miss Arundell's address.' The door opened, I turned round, and judge of my feelings when I beheld Richard! . . . . . We rushed into each other's arms. . . . . We went down-stairs and Richard called a cab, and he put me in and told the man to drive about anywhere. He put his arm round my waist, and I put my head on his shoulder.''[1] Burton had come back more like a mummy than a man, with cadaverous face, brown-yellow skin hanging in bags, his eyes protruding and his lips drawn away from his teeth—the legacy of twenty-one attacks of fever.

(1) *The Romance of Isabel Lady Burton*, i., 149.

When the question of their marriage was brought before her parents, Mr. Arundell not only offered no impediment, but remarked: "I do not know what it is about that man, I cannot get him out of my head. I dream of him every night," but Mrs. Arundell still refused consent. She reiterated her statement that whereas the Arundells were staunch old English Catholics, Burton professed no religion at all, and declared that his conversation and his books proclaimed him an Agnostic. Nor is it surprising that she remained obdurate, seeing that the popular imagination still continued to run riot over his supposed enormities. The midnight hallucinations of De Quincey seemed to be repeating themselves in a whole nation. He had committed crimes worthy of the Borgias. He had done a deed which the ibis and the crocodile trembled at. Miss Arundell boldly defended him against all adversaries; and was for long very bitter against her mother, though she admitted afterwards that, circumstances considered, Mrs. Arundell's opposition was certainly logical.

"As we cannot get your mother's consent," said Burton, "we had better marry without it."

"No," replied Miss Arundell, "that will not do," nor could any argument turn her.

"You and your mother have certainly one characteristic in common," was the comment. "You are as obstinate as mules."

Burton was not without means, for on the death of his father he inherited some £16,000, but he threw his money about with the recklessness of an Aladdin, and 16 million would have gone the same way. It was all, however, or nearly all spent in the service of the public. Every expedition he made, and every book he published left him considerably the poorer. So eager for exploration was he that before the public had the opportunity to read about one expedition, he had started on another. So swiftly did he write, that before one book had left the binders, another was on its way to the printers. Systole, diastole, never ceasing—never even pausing. Miss Arundell being inflexible, Burton resolved to let the matter remain nine months in abeyance, and, inactivity being death to him, he then shot off like a rocket to America. One day in April (1860) Miss Arundell received a brief letter the tenor of which was as follows:—

"I am off to Salt Lake City, and shall be back in December. Think well over our affair, and if your mind is then made up we will marry.

Being the first intimation of his departure—for as usual there had been no good-bye—the message gave her a terrible shock. Hope fled, and a prostrating illness followed. The belief that he would be killed pressed itself upon her and returned with inexplicable insistence. She picked up a newspaper, and the first thing that met her eye was a paragraph headed "Murder of Captain Burton." The shock was terrible, but anxious enquiry revealed the murdered man to be another Captain Burton, not her Richard.

40. Brigham Young. April 1860 to November 1860.

It was natural that, after seeing the Mecca of the Mohammedans, Burton should turn to the Mecca of the Mormons, for he was always attracted by the centres of the various faiths, moreover he wished to learn the truth about a city and a religion that had previously been described only by the biassed. One writer, for instance—a lady—had vilified Mormonism because "some rude men in Salt Lake City had walked over a bridge before her." It was scarcely the most propitious moment to start on such a journey. The country was torn with intestine contentions. The United States Government were fighting the Indians, and the Mormons were busy stalking one another with revolvers. Trifles of this kind, however, did not weigh with Burton. After an uneventful voyage across the Atlantic, and a conventional journey overland, he arrived at St. Joseph, popularly St. Jo, on the Missouri. Here he clothed himself like a backwoodsman, taking care, however, to put among his luggage a silk hat and a frock coat in order to make an impression among the saints. He left St. Jo on August 7th, and at Alcali Lake saw the curious spectacle of an Indian remove. The men were ill-looking, and used vermilion where they ought to have put soap; the squaws and papooses comported with them; but there was one pretty girl who had "large, languishing eyes, and sleek black hair like the ears of a King Charles Spaniel." The Indians followed Burton's waggon for miles, now and then peering into it and crying "How! How!" the normal salutation. His way then lay by darkling canons, rushing streams and stupendous beetling cliffs fringed with pines. Arrived at his destination, he had no difficulty, thanks to the good offices of a fellow traveller, in mixing in the best Mormon Society. He found himself in a Garden City. Every householder had from five to ten acres in the suburbs, and one and a half close at home; and the people seemed happy. He looked in vain, however, for the spires of the Mormon temple which a previous writer had described prettily as glittering in the sunlight. All he could find was "a great hole in the ground," said to be the beginning of a baptismal font, with a plain brick building, the Tabernacle, at a little distance. After a service at the "Tabernacle" he was introduced to Brigham Young, a farmer-like man of 45, who evinced much interest in the Tanganyika journey and discussed stock, agriculture and religion; but when Burton asked to be admitted as a Mormon, Young replied, with a smile, "I think you've done that sort of thing once before, Captain." So Burton was unable to add Mormonism to his five or six other religions. Burton then told with twinkling eyes a pitiful tale of how he, an unmarried man, had come all the way to Salt Lake City, requiring a wife, but had found no wives to be had, all the ladies having been snapped up by the Saints. A little later the two men, who had taken a stroll together, found themselves on an eminence which commanded a view both of the Salt Lake City and the Great Salt Lake. Brigham Young pointed out the various spots of interest, "That's Brother Dash's house, that block just over there is occupied by Brother X's wives. Elder Y's wives reside in the next block and Brother *Z*'s wives in that beyond it. My own wives live in that many-gabled house in the middle."

Waving his right hand towards the vastness of the great Salt Lake, Burton exclaimed, with gravity:

"Water, water, everywhere"

and then waving his left towards the city, he added, pathetically:

"But not a drop to drink."

Brigham Young, who loved a joke as dearly as he loved his seventeen wives, burst out into hearty laughter. In his book, "The City of the Saints," Burton assures us that polygamy was admirably suited for the Mormons, and he gives the religious, physiological and social motives for a plurality of wives then urged by that people. Economy, he tells us, was one of them. "Servants are rare and costly; it is cheaper and more comfortable to marry them. Many converts are attracted by the prospect of becoming wives, especially from places like Clifton, near Bristol, where there are 64 females to 36 males. The old maid is, as she ought to be, an unknown entity."(1)

Burton himself received at least one proposal of marriage there; and the lady, being refused, spread the rumour that it was the other way about. "Why," said Burton, "it's like

A certain Miss Baxter,
Who refused a man before he'd axed her." (2)

As regards the country itself nothing struck him so much as its analogy to Palestine. A small river runs from the Wahsatch Mountains, corresponding to Lebanon, and

(1) He is, of course, simply endorsing the statement of Hippocrates *De Genitura*: "Women, if married, are more healthy, if not, less so."

(2) The anecdotes in this chapter were told me by one of Burton's friends. They are not in his books.

flows into Lake Utah, which represents Lake Tiberias, whence a river called the Jordan flows past Salt Lake City into the Great Salt Lake, just as the Palestine Jordan flows into the Dead Sea.

From Salt Lake City, Burton journeyed by coach and rail to San Francisco, whence he returned home via Panama.

41. Marriage. 22nd January 1861.

He arrived in England at Christmas 1860, and Miss Arundell, although her mother still frowned, now consented to the marriage. She was 30 years old, she said, and could no longer be treated as a child. Ten years had elapsed since Burton, who was now 40, had first become acquainted with her, and few courtships could have been more chequered.

"I regret that I am bringing you no money," observed Miss Arundell.

"That is not a disadvantage as far as I am concerned," replied Burton "for heiresses always expect to lord it over their lords."—"We will have no show," he continued, "for a grand marriage ceremony is a barbarous and an indelicate exhibition." So the wedding, which took place at the Bavarian Catholic Church, Warwick Street, London, on 22nd January 1861, was all simplicity. As they left the church Mrs. Burton called to mind Gipsy Hagar, her couched eyes and her reiterated prophecy. The luncheon was spread at the house of a medical friend, Dr. Bird, 49, Welbeck Street, and in the midst of it Burton told some grisly tales of his adventures in the Nedj and Somaliland, including an account of the fight at Berbera.

"Now, Burton," interrupted Dr. Bird, "tell me how you feel when you have killed a man." To which Burton replied promptly and with a sly look, "Quite jolly, doctor! how do you?" After the luncheon Burton and his wife walked down to their lodgings in Bury Street, St. James's, where Mrs. Burton's boxes had been despatched in a four-wheeler; and from Bury Street, Burton, as soon as he could pick up a pen, wrote in his fine, delicate hand as follows to Mr. Arundell:

"January 23rd 1861,[1]
"Bury Street,
"St. James.

"My dear Father,

"I have committed a highway robbery by marrying your daughter Isabel, at Warwick Street Church, and before the Registrar—the details she is writing to her mother

"It only remains to me to say that I have no ties or liaisons of any sort, that the marriage is perfectly legal and respectable. I want no money with Isabel: I can work, and it will be my care that Time shall bring you nothing to regret.

"I am
"Yours sincerely,
"RICHARD F. BURTON."

"There is one thing," said Burton to his wife, "I cannot do, and that is, face congratulations, so, if you are agreeable, we will pretend that we have been married some months." Such matters, however, are not easy to conceal, and the news leaked out. "I am surprised," said his cousin, Dr. Edward J. Burton, to him a few days later, "to find that you are married." "I am myself even more surprised than you," was the reply. "Isabel is a strong-willed woman. She was determined to have her way and she's got it."

With Mr. Arundell, Burton speedily became a prime favourite, and his attitude towards his daughter was Metastasio's:

"Yes, love him, love him,
He is deserving even of such infinite bliss;"

but Mrs. Arundell, poor lady, found it hard to conquer her prejudice. Only a few weeks before her death she was heard to exclaim, "Dick Burton is no relation of mine."

(1) This letter was given by Mrs. FitzGerald (Lady Burton's sister) to Mr. Foskett of Camberwell. It is now in the library there, and I have to thank the library committee for the use of it.

Let us charitably assume, however, that it was only in a moment of irritation. Isabel Burton, though of larger build than most women, was still a dream of beauty; and her joy in finding herself united to the man she loved gave her a new radiance. Her beauty, however, was of a rather coarse grain, and even those most attached to her remarked in her a certain lack of refinement. She was a goddess at a little distance.

Her admiration of her husband approached worship. She says, "I used to like to sit and look at him; and to think 'You are mine, and there is no man on earth the least like you.'" Their married life was not without its jars, but a more devoted wife Burton could not have found; and he, though certainly in his own fashion, was sincerely and continuously attached to her. If the difference in their religious opinions sometimes led to amusing skirmishes, it was, on the other hand, never allowed to be a serious difficulty. The religious question, however, often made unpleasantness between Mrs. Burton and Lady Stisted and her daughters—who were staunch Protestants of the Georgian and unyielding school. When the old English Catholic and the old English Protestant met there were generally sparks. The trouble originated partly from Mrs. Burton's impulsiveness and want of tact. She could not help dragging in her religion at all sorts of unseasonable times. She would introduce into her conversation and letters remarks that a moment's reflection would have told her could only nauseate her Protestant friends. "The Blessed Virgin," or some holy saint or other was always intruding on the text. Her head was lost in her heart. She was once in terrible distress because she had mislaid some trifle that had been touched by the Pope, though not in more distress, perhaps, than her husband would have been had he lost his sapphire talisman, and she was most careful to see that the lamps which she lighted before the images of certain saints never went out. Burton himself looked upon all this with amused complacency and observed that she was a figure strayed somehow from the Middle Ages. If the mediæval Mrs. Burton liked to illuminate the day with lamps or camphorated tapers, that, he said, was her business; adding that the light of the sun was good enough for him. He objected at first to her going to confession, but subsequently made no further reference to the subject. Once, even, in a moment of weakness, he gave her five pounds to have masses said for her dead brother; just as one might give a child a penny to buy a top. He believed in God, and tried to do what he thought right, fair and honourable, not for the sake of reward, as he used to say, but simply because it was right, fair and honourable. Occasionally he accompanied his wife to mass, and she mentions that he always bowed his head at "Hallowed be Thy Name," which "shows," as Dr. Johnson would have commented, "that he had good principles." Mrs. Burton generally called her husband "Dick," but frequently, especially in letters, he is "The Bird," a name which he deserved, if only on account of his roving propensities. Often, however, for no reason at all, she called him "Jimmy," and she was apt in her admiration of him and pride of possession, to Dick and Jimmy it too lavishly among casual acquaintances. Indeed, the tyranny of her heart over her head will force itself upon our notice at every turn. It is pleasant to be able to state that Mrs. Burton and Burton's "dear Louisa" (Mrs. Segrave) continued to be the best of friends, and had many a hearty laugh over bygone petty jealousies. One day, after calling on Mrs. Segrave, Burton and his wife, who was dressed in unusual style, lunched with Dr. and Mrs. E. J. Burton. "Isabel looks very smart to-day," observed Mrs. E. J. Burton. "Yes," followed Burton, "she always wears her best when we go to see my dear Louisa."

Burton took a pleasure in sitting up late. "Indeed," says one of his friends, "he would talk all night in preference to going to bed, and, in the Chaucerian style, he was a brilliant conversationalist, and his laugh was like

THE "WEDDING PORTRAITS" OF CAPTAIN AND MRS. BURTON.

FROM THE ORIGINAL IN THE CENTRAL LIBRARY, AT CAMBERWELL. *(By permission of the Committee.)*

Photo by Albert Flint, Camberwell.

the rattle of a pebble across a frozen pond." "No man of sense," Burton used to say, "rises, except in mid-summer, before the world is brushed and broomed, aired and sunned." Later, however, he changed his mind, and for the last twenty years of his life he was a very early riser.

Among Burton's wedding gifts were two portraits—himself and his wife—in one frame, the work of Louis Desanges, the battle painter whose acquaintance he had made when a youth at Lucca. Burton appears with with Atlantean shoulders, strong mouth, penthouse eye-brows, and a pair of enormous pendulous moustaches, which made him look very like a Chinaman. Nor was this an accident, for his admiration of the Chinese was always intense. He regarded them as "the future race of the East," just as he regarded the Slav as the future race of Europe. Many years later he remarked of Gordon's troops, that they had shown the might that was slumbering in a nation of three hundred millions. China armed would be a colossus. Some day Russia would meet China face to face—the splendid empire of Central Asia the prize. The future might of Japan he did not foresee.

Says Lady Burton: "We had a glorious season, and took up our position in Society. Lord Houghton (Monckton Milnes) was very much attached to Richard, and he settled the question of our position by asking his friend, Lord Palmerston, to give a party, and to let me be the bride of the evening, and when I arrived Lord Palmerston gave me his arm. . . . Lady Russell presented me at Court 'on my marriage.'"[1]

Mrs. Burton's gaslight beauty made her the cynosure of all eyes.

At Fryston, Lord Houghton's seat, the Burtons met Carlyle, Froude, Mr. A. C. Swinburne, who had just

(1) *Life*, i., 345.

published his first book, *The Queen Mother and Rosamund*,[1] and Vambery, the Hungarian linguist and traveller. Born in Hungary, of poor Jewish parents, Vambery had for years a fierce struggle with poverty. Having found his way to Constantinople, he applied himself to the study of Oriental languages, and at the time he visted Fryston he was planning the most picturesque event of his life—namely, his journey to Khiva, Bokhara and Samarcand, which in emulation of Burton he accomplished in the disguise of a dervish.[2] He told the company some Hungarian tales and then Burton, seated cross-legged on a cushion, recited portions of FitzGerald's adaptation of Omar Khayyam,[3] the merits of which he was one of the first to recognise. Burton and Lord Houghton also met frequently in London, and they corresponded regularly for many years.[4] "Richard and I," says Mrs. Burton, writing to Lord Houghton 12th August 1874, "would have remained very much in the background if you had not taken us by the hand and pulled us into notice." A friendship also sprang up between Burton and Mr. Swinburne, and the Burtons were often the guests of Mrs. Burton's uncle, Lord Gerard, who resided at Garswood, near St. Helens, Lancashire.

42. At Lord Houghton's.

(1) 1861

(2) Vambery's work, *The Story of my Struggles*, appeared in October 1904.

(3) The first edition appeared in 1859. Burton's works contain scores of allusions to it. *To the Gold Coast*, ii., 164. *Arabian Nights* (many places), etc., etc.

(4) *Life of Lord Houghton*, ii., 300.

LADY STISTED.

FROM AN OIL PAINTING IN THE POSSESSION OF MR. E. S. MOSTYN PRYCE, GUNLEY HALL, STAFFORDSHIRE.

Photo by J. H. Anderson & Son, Welshpool.

CHAPTER XI

AUGUST 1861—NOVEMBER 1863

FERNANDO PO

BIBLIOGRAPHY:

18. Wanderings in West Africa. 2 vols. 1863.
19. Prairie Traveller, by R. B. Marcy. Edited by Burton 1863.
20. Abeokuta and the Cameroons. 2 vols. 1863.
21. A Day among the Fans. 17th February 1863.
22. The Nile Basin, 1864.

43. African Gold.

As the result of his exceptional services to the public Burton had hoped that he would obtain some substantial reward; and his wife persistently used all the influence at her disposal to this end. Everyone admitted his immense brain power, but those mysterious rumours due to his enquiries concerning secret Eastern habits and customs dogged him like some terrible demon. People refused to recognise that he had pursued his studies in the interest of learning and science. They said, absurdly enough, "A man who studies vice must be vicious." His insubordination at various times, his ungovernable temper, and his habit of saying out bluntly precisely what he thought, also told against him. Then did Mrs. Burton commence that great campaign which is her chief title to fame—the defence of her husband. Though, as we have already shown, a person of but superficial education; though, life through, she never got more than a smattering of any one branch of knowledge; nevertheless by dint of unremitting effort she eventually prevailed upon the public to regard Burton with her own eyes. She wrote letters to friends, to enemies, to the press. She wheedled, she bullied, she threatened, she took a hundred other courses—all with one purpose. She was very often woefully indiscreet, but nobody can withhold admiration for her. Burton was scarcely a model husband—he was too peremptory and inattentive for that—but this self-sacrifice and hero worship naturally told on him, and he became every year more deeply grateful to her. He laughed at her foibles—he twitted her on her religion and her faulty English, but he came to value the beauty of her disposition, and the goodness of her heart even more highly than the graces of her person. All, however, that his applications, her exertions, and the exertions of her friends could obtain from the Foreign Secretary (Lord Russell)[1] was the Consulship of that white man's grave, Fernando Po, with a salary of £700 a year. In other words he was civilly shelved to a place where all his energies would be required for keeping himself alive. "They want me to die," said Burton, bitterly, "but I intend tto live, just to spite the devils." It is the old tale, England breeds great men, but grudges them opportunities for the manifestation of their greatness.

The days that remained before his departure, Burton spent at various Society gatherings, but the pleasures participated in by him and his wife were neutralised by a great disaster, namely the loss of all his Persian and Arabic manuscripts in a fire at Grindley's where they had been stored. He certainly took his loss philosophically; but he could never think of the event without a sigh.

Owing to the unwholesomeness of the climate of Fernando Po, Mrs. Burton was, of course, unable to accompany him. They separated at Liverpool, 24th August 1861. An embrace, "a heart wrench;" and then a wave of the handkerchief, while "the Blackbird" African steam ship fussed its way out of the Mersey, having on board the British scape-goat sent away—"by the hand

(1) Lord Russell was Foreign Secretary from 1859-1865.

CAPTAIN BURTON.

FROM A PHOTOGRAPH TAKEN IN 1862.

of a fit man "— one "Captain English "— into the wilderness of Fernando Po. "Unhappily," commented Burton, "I am not one of those independents who can say *ce n'est que le premier pas qui coûte.*" The stoic, however, after a fair fight, eventually vanquished the husband. Still he did not forget his wife; and in his *Wanderings in West Africa*, a record of this voyage, there is a very pretty compliment to her which, however, only the initiated would recognise. After speaking of the black-haired, black-eyed women of the South of Europe, and giving them their due, he says, "but after a course of such charms, one falls back with pleasure upon brown, yellow, or what is better than all, red-auburn locks and eyes of soft, limpid blue." How the blue eyes of Mrs. Burton must have glistened when she read those words; and we can imagine her taking one more look in the glass to see if her hair really was red-auburn, as, of course, it was.

Burton dedicated this work to the "True Friends" of the Dark Continent, "not to the 'Philanthropist' or to Exeter Hall." (1) One of its objects was to give a trustworthy account of the negro character and to point out the many mistakes that well-intentioned Englishmen had made in dealing with it. To put it briefly, he says that the negro (2) is an inferior race, and that neither education nor anything else can raise it to the level of the white. After witnessing, at the Grand Bonny River, a horrid exhibition called a Juju or sacrifice house, he wrote, "There is apparently in this people [the negroes] a physical delight in cruelty to beast as well as to man. The sight of suffering seems to bring them an enjoyment without which the world is tame; probably the wholesale murderers and torturers of history, from Phalaris and Nero downwards, took an animal and sensual pleasure in the look of blood, and in the inspection of mortal agonies. I can see no other explanation of the

(1) *Wanderings in West Africa.* 2 vols., 1863.

(2) The genuine black, not the mulatto, as he is careful to point out. Elsewhere he says the negro is always eight years old—his mind never develops. *Mission to Gelele*, i, 216.

phenomena which meet my eye in Africa. In almost all the towns on the Oil Rivers, you see dead or dying animals in some agonizing position." (1)

Cowper had written :

> " Skins may differ, but affection
> Dwells in white and black the same ; "

"which I deny," comments Burton, "affection, like love, is the fruit of animalism refined by sentiment." He further declares that the Black is in point of affection inferior to the brutes. "No humane Englishman would sell his dog to a negro." (2) The phrase "God's image in ebony" lashed him to fury.

Of his landing at Sierra Leone he gives the following anecdote: (3) "The next day was Sunday, and in the morning I had a valise carried up to the house to which I had been invited. When I offered the man sixpence, the ordinary fee, he demanded an extra sixpence, 'for breaking the Sabbath.' I gave it readily, and was pleased to find that the labours of our missionaries had not been in vain." At Cape Coast Castle, he recalled the sad fate of "L.E.L." (4) and watched the women "panning the sand of the shore for gold." He found that, in the hill region to the north, gold digging was carried on to a considerable extent. "The pits," he says, "varying from two to three feet in diameter, and from twelve to fifty feet deep, are often so near the roads that loss of life has been the result. Shoring up being little known, the miners are not infrequently buried alive. . . . This Ophir, this California, where every river is a Tmolus and a Pactolus, every hillock is a gold-field—does not contain a cradle, a puddling-machine, a quartz crusher,

(1) *Wanderings in West Africa*, vol. ii., p. 283.

(2) See *Mission to Gelele*, ii., 126.

(3) Although the anecdote appears in his *Abeokuta* it seems to belong to this visit.

(4) Mrs. Maclean, "L.E.L.," went out with her husband, who was Governor of Cape Coast Castle. She was found poisoned 15th October 1838, two days after her arrival. Her last letters are given in *The Gentleman's Magazine*, February 1839.

a pound of mercury." That a land apparently so wealthy should be entirely neglected by British capitalists caused Burton infinite surprise, but he felt certain that it had a wonderful future. His thoughts often reverted thither, and we shall find him later in life taking part in an expedition sent out to report upon certain of its gold fields.[1]

By September 26th the "Blackbird" lay in Clarence Cove, Fernando Po; and the first night he spent on shore, Burton, whose spirits fell, wondered whether he was to find a grave there like that other great African traveller, the Cornish Richard Lander.[2]

Fernando Po,[3] he tell us, is an island in which man finds it hard to live and very easy to die. It has two aspects. About Christmas time it is "in a state deeper than rest: 44. Anecdotes.

"A kind of sleepy Venus seemed Dudu."

But from May to November it is the rainy season. The rain comes down "a sheet of solid water, and often there is lightning accompanied by deafening peals of thunder." The capital, Sta. Isabel, née Clarence, did not prepossess him. Pallid men—chiefly Spaniards—sat or lolled languidly in their verandahs, or crawled about the baking-hot streets. Strangers fled the place like a pestilence. Fortunately the Spanish colony were just establishing a Sanitarium—Sta. Cecilia—400 mètres above sea level; consequently health was within reach of those who would take the trouble to seek it; and Burton was not slow to make a sanitarium of his own even higher up. To the genuine natives or Bubés he was distinctly attracted. They lived in sheds without walls, and wore nothing except a hat, which prevented the tree snakes from falling on them. The impudence of the

(1) See Chapter xxii.

(2) Lander died at Fernando Po, 16th February 1834.

(3) For notes on Fernando Po see Laird and Oldfield's *Narrative of an Expedition into the Interior of Africa*, etc. (1837), Winwood Reade's *Savage Africa*, and Rev. Henry Roe's *West African Scenes* (1874).

negroes, however, who would persist in treating the white man not even as an equal, but as an inferior, he found to be intolerable. Shortly after his arrival "a nigger dandy" swaggered into the consulate, slapped him on the back in a familiar manner, and said, with a loud guffaw, "Shake hands, consul. How d'ye do?" Burton looked steadily at the man for a few moments, and then calling to his canoe-men said, "Hi, Kroo-boys, just throw this nigger out of window, will you?" The boys, delighted with the task, seized the black gentleman by his head and feet, and out of the window he flew. As the scene was enacted on the ground floor the fall was no great one, but it was remarked that henceforward the niggers of Fernando Po were less condescending to the Consul. When night fell and the fire-flies began to glitter in the orange trees, Burton used to place on the table before him a bottle of brandy, a box of cigars, and a bowl containing water and a handkerchief and then write till he was weary;[1] rising now and again to wet his forehead with the handkerchief or to gaze outside at the palm plumes, transmuted by the sheen of the moon into lucent silver—upon a scene that would have baffled the pen even of an Isaiah or a Virgil.

The captains of ships calling at Sta. Isabel were, it seems, in the habit of discharging their cargoes swiftly and steaming off again without losing a moment. As this caused both inconvenience and loss to the merchants from its allowing insufficient time to read and answer correspondence, they applied to Burton for remedy. After the next ship had discharged, its captain walked into the Consulate and exclaimed off-handedly, "Now, Consul, quick with my papers; I want to be off." Burton looked up and replied unconcernedly: "I haven't finished my letters." "Oh d—— your letters," cried the captain, "I can't wait for them." "Stop a bit," cried Burton, "let's refer to your contract," and he unfolded the paper. "According to this, you have to stay here eighteen

(1) Told me by the Rev. Henry Roe.

hours' daylight, in order to give the merchants an opportunity of attending to their correspondence." "Yes," followed the captain, "but that rule has never been enforced." "Are you going to stay?" enquired Burton. "No," replied the captain, with an oath. "Very good," followed Burton. "Now I am going straight to the governor's and I shall fire two guns. If you go one minute before the prescribed time expires I shall send the first shot right across your bows, and the second slap into you. Good-day."(1) The captain did not venture to test the threat; and the merchants had henceforth no further trouble under this head.

During his Consulship, Burton visited a number of interesting spots on the adjoining African coast, including Abeokuta(2) and Benin, but no place attracted him more than the Cameroon country; and his work "*Two Trips to Gorilla Land*(3) is one of the brightest and raciest of his books. The Fan cannibals seem to have specially fascinated him. "The Fan," he says "like all inner African tribes, with whom fighting is our fox-hunting, live in a chronic state of ten days' war. Battles are not bloody; after two or three warriors have fallen their corpses are dragged away to be devoured, their friends save themselves by flight, and the weaker side secures peace by paying sheep and goats." Burton, who was present at a solemn dance led by the king's eldest daughter, Gondebiza, noticed that the men were tall and upright, the women short and stout. On being addressed "Mbolane," he politely replied "An," which in cannibal-land is considered good form. He could not, however, bring himself to admire Gondebiza, though the Monsieur Worth of Fanland had done his utmost for her. Still, she must have looked really engaging in a thin

45. Fans and Gorillas.

(1) *Life*, and various other works.

(2) See *Abeokuta and the Cameroons*, 2 vols., 1863.

(3) *Two Trips to Gorilla Land and the Cataracts of the Congo*, 2 vols., 1876.

pattern of tattoo, a gauze work of oil and camwood, a dwarf pigeon tail of fan palm for an apron, and copper bracelets and anklets. The much talked of gorilla Burton found to be a less formidable creature than previous travellers had reported. "The gorilla," he, says, in his matter-of-fact way, "is a poor devil ape, not a hellish dream creature, half man, half beast." Burton not only did not die at Fernando Po, he was not even ill. Whenever langour and fever threatened he promptly winged his way to his eyrie on the Pico de Sta. Isabel, where he made himself comfortable and listened with complaisance to Lord Russell and friends three thousand miles away fuming and gnashing their teeth.

46. The Anthropological Society, 6th Jan. 1863.

After an absence of a year and a half, Burton, as the result of his wife's solicitation at the Foreign Office, obtained four months' leave. He reached England in December 1862 and spent Christmas with her at Wardour Castle, the seat of her kinsman, Lord Arundell. His mind ran continually on the Gold Coast and its treasures. "If you will make me Governor of the Gold Coast," he wrote to Lord Russell, "I will send home a million a year," but in reply, Russell, with eyes unbewitched (1) observed caustically that gold was getting too common. Burton's comment was an explosion that terrorised everyone near him. He then amused himself by compiling a pamphlet on West African proverbs, one of which, picked up in the Yorubas country, ran, oddly enough: "Anger draweth arrows from the quiver: good words draw kolas from the bag."

The principal event of this holiday was the foundation, with the assistance of Dr. James Hunt, of the Anthropological Society of London (6th January 1863). The number who met was eleven. Says Burton, "Each had his own doubts and hopes and fears touching the vitality of the new-born. Still, we knew that our cause was

(1) "Who first bewitched our eyes with Guinea gold." Dryden, *Annus Mirabilis*, 67.

good. . . . We all felt the weight of a great want. As a traveller and a writer of travels I have found it impossible to publish those questions of social economy and those physiological observations, always interesting to our common humanity, and at times so valuable." The *Memoirs of the Anthropological Society*,(1) met this difficulty. Burton was the first president, and in two years the Society, which met at No. 4, St. Martin's Place, had 500 members. "These rooms," Burton afterwards commented, "now offer a refuge to destitute truth. There any man, monogenist, polygenist, eugenestic or dysgenestic, may state the truth as far as is in him." The history of the Society may be summed up in a few words. In 1871 it united with the Ethnological Society and formed the Anthropological Institute of Great Britain. In 1873 certain members of the old society, including Burton, founded the London Anthropological Society, and issued a periodical called *Anthropologia*, of which Burton wrote in 1885, "My motive was to supply travellers with an organ which would rescue their observations from the outer darkness of manuscript and print their curious information on social and sexual matters out of place in the popular book intended for the Nipptisch, and indeed better kept from public view. But hardly had we begun when 'Respectability,' that whited sepulchre full of all uncleanness, rose up against us. 'Propriety' cried us down with her brazen, blatant voice, and the weak-kneed brethren fell away.(2) Yet the organ was much wanted and is wanted still."(3) Soon after the founding of the Society Burton, accompanied by his wife, took a trip to Madeira and then proceeded to Teneriffe, where they parted, he going on to Fernando Po and she returning to England; but during the next few years she made several journeys to Teneriffe, where, by arrangement, they periodically met.

(1) Incorporated subsequently with a Quarterly Journal, *The Anthropological Review*.

(2) See Chapter xxix., § 140.

(3) Foreword to *The Arabian Nights*, vol. 1. *The Arabian Nights*, of course, was made to answer the purpose of this organ.

CHAPTER XII

29TH NOVEMBER 1863 TO 15TH SEPTEMBER 1865

GELELE

BIBLIOGRAPHY:

23. A Mission to the King of Dahomé. 2 vols., 1864.
24. Notes on Marcy's Prairie Traveller. Anthropological Review, 1864.

47. Whydah and its Deity. 29th November 1863.

In November 1863 the welcome intelligence reached Burton that the British Government had appointed him commissioner and bearer of a message to Gelele, King of Dahomey. He was to take presents from Queen Victoria and to endeavour to induce Gelele to discontinue both human sacrifices and the sale of slaves. Mrs. Burton sadly wanted to accompany him. She thought that with a magic lantern and some slides representing New Testament scenes she could convert Gelele and his court from Fetishism to Catholicism.[1] But Burton, who was quite sure that he could get on better alone, objected that her lantern would probably be regarded as a work of magic, and that consequently both he and she would run the risk of being put to death for witchcraft. So, very reluctantly, she abandoned the idea. Burton left Fernando Po in the "Antelope" on 29th November 1863, and, on account of the importance attached by savages to pageantry, entered Whydah, the port of Dahomey, in some state. While waiting for the royal permit to start up country he amused himself by looking round the town. Its lions were the Great Market and

(1) See *Wanderings in West Africa*, vol. 2, p. 91, footnote.

the Boa Temple. The latter was a small mud hut, with a thatched roof; and of the 'boas,' which turned out to be pythons, he counted seven, each about five feet long. The most popular deity of Whydah, however, was the Priapic Legba, a horrid mass of red clay moulded into an imitation man with the abnormalities of the Roman deity. "The figure," he tells us, "is squat, crouched, as it were, before its own attributes, with arms longer than a gorilla's. The head is of mud or wood rising conically to an almost pointed poll; a dab of clay represents the nose; the mouth is a gash from ear to ear. This deity almost fills a temple of dwarf thatch, open at the sides. . . . Legba is of either sex, but rarely feminine. . . . In this point Legba differs from the classical Pan and Priapus, but the idea involved is the same. The Dahoman, like almost all semi-barbarians, considers a numerous family the highest blessing." The peculiar worship of Legba consisted of propitiating his or her characteristics by unctions of palm oil, and near every native door stood a clay Legba-pot of cooked maize and palm oil, which got eaten by the turkey-buzzard or vulture. This loathsome fowl, perched upon the topmost stick of a blasted calabash tree, struck Burton as the most appropriate emblem of rotten and hopeless Dahomey.

48. The Amazons.

Gelele's permit having arrived, the mission lost no time in proceeding northward. Burton was accompanied by Dr. Cruikshank of the "Antelope," a coloured Wesleyan minister of Whydah, named Bernisco, and a hundred servants. At every halting place the natives capered before them and tabored a welcome, while at Kama, where Gelele was staying, they not only played, but burst out with an extemporaneous couplet in Burton's honour:

> "Batunu[1] he hath seen the world with its kings and caboceers,
> He now cometh to Dahomey, and he shall see everything here."

Burton presently caught sight of Gelele's body-guard

(1) Burton.

of 1,000 women—the famous Amazons, who were armed with muskets, and habited in tunics and white calottes. With great protruding lips, and no chin to speak of, they were surely the ugliest women in the world. Of their strength, however, there was no question, and Burton says that all the women of Dahomey are physically superior to the men, which accounts for the employment of so many of them as soldiers. The Amazons were bound to celibacy, and they adhered to it so scrupulously that when Burton arrived, there were only 150 under confinement for breaking their vow. Gelele who was 45 years of age, and six feet high, sat under the shade of a shed-gate, smoking a pipe, with a throng of his wives squatted in a semi-circle round him. All were ugly to a wonder, but they atoned for their deplorable looks by their extreme devotion to, or rather adulation of their master. When perspiration appeared upon the royal brow, one of them at once removed it with the softest cloth, if his dress was disarranged it was instantly adjusted, when he drank every lip uttered an exclamation of blessing. Gelele, drowsy with incense, received Burton kindly, and treated him during the whole of his stay with hospitality. He also made some display of pageantry, though it was but a tawdry show. At the capital, Abomey, "Batunu" was housed with a salacious old "Afa-diviner"[1] called Buko-no, who was perpetually begging for aphrodisiacs.

Upon Gelele's arrival at Abomey the presents from the Queen were delivered; and on December 28th what was called "The Customs" began, that is the slaughtering of criminals and persons captured in war. Burton begged off some of the victims, and he declared that he would turn back at once if any person was killed before his eyes. He tells us, however, that in the case of the King of Dahomey, human sacrifice is not attributable to cruelty. "It is a touching instance of the King's filial piety, deplorably

49. "The Customs."

(1) Afa is the messenger of fetishes and of deceased friends. Thus by the Afa diviner people communicate with the dead.

mistaken, but perfectly sincere." The world to come is called by the Dahomans "Deadland." It receives the 'nidon' or soul; but in "Deadland" there are no rewards or punishments. Kings here are kings there, the slave is a slave for ever and ever; and people occupy themselves just the same as on earth. As the Dahoman sovereign is obliged to enter Deadland, his pious successor takes care that the deceased shall make this entrance in royal state, "accompanied by a ghostly court of leopard wives, head wives, birthday wives, Afa wives, eunuchs, singers, drummers, bards and soldiers." Consequently when a king dies some 500 persons are put to death, their cries being drowned by the clangour of drums and cymbals. This is called the "Grand Customs." Every year, moreover, decorum exacts that the firstfruits of war and all criminals should be sent as recruits to swell the king's retinue. Hence the ordinary "Annual Customs," at which some 80 perish. Burton thus describes the horrors of the approach to the "palace"—that is to say, a great thatched shed—on the fifth day of the "Customs." "Four corpses, attired in their criminal's shirts and night-caps, were sitting in pairs upon Gold Coast stools, supported by a double-storied scaffold, about forty feet high, of rough beams, two perpendiculars and as many connecting horizontals. At a little distance on a similar erection, but made for half the number, were two victims, one above the other. Between these substantial structures was a gallows of thin posts, some thirty feet tall, with a single victim hanging by the heels head downwards." Hard by were two others dangling side by side. The corpses were nude and the vultures were preying upon them, and squabbling over their hideous repast. All this was grisly enough, but there was no preventing it. Then came the Court revels. The king danced in public, and at his request, Burton and Dr. Cruikshank also favoured the company. Bernisco, when called upon, produced a concertina and played "O, let us be joyful, when we meet to part no more." The idea, however, of getting to any place where he would never be separated from Gelele, his brutish court, his corpses and his vultures severely tried Burton's gravity. Gelele, who was preparing for an unprovoked attack upon Abeokuta, the capital of the neighbouring state of Lagos, now made some grandiose and rhapsodical war speeches and spoke vauntingly of the deeds that he and his warriors meant to perform, while every now and then the younger bloods, eager to flesh their spears, burst out with:

> "When we go to war we must slay men,
> And so must Abeokuta be destroyed."

The leave-taking between Gelele and "Batunu" was affecting. Burton presented his host with a few not very valuable presents, and Gelele in return pressed upon his guest a cheap counterpane and a slave boy who promptly absconded.

Whydah was reached again on 18th February 1864, and within a week came news that Gelele, puffed up with confidence and vainglory, had set out for Abeokuta, and was harrying that district. He and his Amazons, however, being thoroughly defeated before the walls of the town, had to return home in what to any other power would have been utter disgrace. They manage things differently, however, in Dahomey, for Gelele during his retreat purchased a number of slaves, and re-entered his capital a triumphing conqueror. Burton considered Gelele, despite his butcherings and vapourings, as, on the whole, quite a phœnix for an African. Indeed, some months after his mission, in conversation with Froude, the historian, he became even warm when speaking of the lenity, benevolence and enlightenment of this excellent king. Froude naturally enquired why, if the king was so benevolent, he did not alter the murderous "Customs." Burton looked up with astonishment. "Alter the Customs!" he said, "Would you have the Archbishop of Canterbury alter the Liturgy!"

To a friend who observed that the customs of Dahomey

were very shocking, Burton replied : " Not more so than those of England."

" But you admit yourself that eighty persons are sacrificed every year."

" True, and the number of deaths in England caused by the crinoline alone numbers 72.[1]

In August 1864 Burton again obtained a few months' leave, and before the end of the month he arrived at Liverpool. It will be remembered that after the Burton and Speke Expedition of 1860 Speke was to go out to Africa again in company with Captain J. Grant.

50. Death of Speke, 15th September 1864.

The expedition not only explored the western and northern shores of the Victoria Nyanza, but followed for some distance the river proceeding northwards from it, which they held, and as we now know, correctly, to be the main stream of the Nile. Burton, however, was still of opinion that the honour of being the head waters of that river belonged to Tanganyika and its affluents. The subject excited considerable public interest and it was arranged that at the approaching Bath meeting of the British Association, Speke and Burton should hold a public disputation upon the great question. Speke's attitude towards Burton in respect to their various discoveries had all along been incapable of defence, while Burton throughout had exhibited noble magnanimity. For example, he had written on 27th June 1863 from the Bonny River to Staff-Commander C. George, " Please let me hear all details about Captain Speke's discovery. He has performed a magnificent feat and now rises at once to the first rank amongst the explorers of the day."[2] Though estranged, the two travellers still occasionally communicated, addressing each other, however, not as " Dear Dick " and " Dear Jack " as aforetime—using, indeed, not " Dear " at all,

(1) This was Dr. Lancaster's computation.

(2) Communicated to me by Mr. W. H. George, son of Staff-Commander C. George, Royal Navy.

but the icy "Sir." Seeing that on public occasions Speke still continued to talk vaingloriously and to do all in his power to belittle the work of his old chief, Burton was naturally incensed, and the disputation promised to be a stormy one. The great day arrived, and no melodramatic author could have contrived a more startling, a more shocking denouement. Burton, notes in hand, stood on the platform, facing the great audience, his brain heavy with arguments and bursting with sesquipedalian and sledge-hammer words to pulverize his exasperating opponent. Mrs. Burton, who had dressed with unusual care, occupied a seat on the platform. "From the time I went in to the time I came out," says one who was present, "I could do nothing but admire her. I was dazed by her beauty." The Council and other speakers filed in. The audience waited expectant. To Burton's surprise Speke was not there. Silence having been obtained, the President advanced and made the thrilling announcement that Speke was dead. He had accidentally shot himself that very morning when out rabbiting.

Burton sank into a chair, and the workings of his face revealed the terrible emotion he was controlling and the shock he had received. When he got home he wept like a child. At this point the grotesque trenches on the tragic. On recovering his calmness, Burton expressed his opinion, and afterwards circulated it, that Speke had committed suicide in order to avoid "the exposure of his misstatements in regard to the Nile sources." In other words, that Speke had destroyed himself lest arguments, subsequently proved to be fundamentally correct, should be refuted. But it was eminently characteristic of Burton to make statements which rested upon insufficient evidence, and we shall notice it over and over again in his career. That was one of the glorious man's most noticeable failings. It would here, perhaps, be well to make a brief reference to the expeditions that settled once and for ever the questions about Tanganyika and the Nile. In March 1870, Henry M. Stanley set out from Bagamoyo in search of Livingstone, whom he found at Ujiji. They spent the early months of 1872 together exploring the north end of Tanganyika, and proved conclusively that the lake had no connection with the Nile basin. In March 1873, Lieutenant Verney Lovett Cameron, who was appointed to the command of an expedition to relieve Livingstone, arrived at Unyanyembe, where he met Livingstone's followers bearing their master's remains to the coast. Cameron then proceeded to Ujiji, explored Tanganyika and satisfied himself that this lake was connected with the Congo system. He then continued his way across the continent and came out at Banguelo, after a journey which had occupied two years and eight months. Stanley, who, in 1874, made his famous journey from Bagamoyo via Victoria Nyanza to Tanganyika and then followed the Congo from Nyangwé, on the Lualaba, to the sea, verified Cameron's conjecture.

At the end of the year 1864 the Burtons made the acquaintance of the African traveller Winwood Reade; and we next hear of a visit to Ireland, which included a day at Tuam, where "the name of Burton was big," on account of the Rector and the Bishop,[1] Burton's grandfather and uncle.

(1) Rev. Edward Burton, Burton's grandfather, was Rector of Tuam. Bishop Burton, of Killala, was the Rev. Edward Burton's brother.

CHAPTER XIII

SEPTEMBER 1865—OCTOBER 1869

SANTOS : BURTON'S SECOND CONSULATE

BIBLIOGRAPHY :

25. Speech before the Anthropological Society. 4th April 1865.
26. Wit and Wisdom from West Africa. 1865.
27. Pictorial Pilgrimage to Mecca and Medina.
28. Psychic Facts, by Francis Baker (Burton). 1865.
29. Notes. . . connected with the Dahoman. 1865.
30. On an Hermaphrodite. 1866.
31. Exploration of the Highlands of the Brazil. 2 vols. 1869.

51. To Santos.

Owing mainly to Mrs. Burton's solicitation, Burton was now transferred from Fernando Po to Santos, in Brazil, so it was no longer necessary for him and his wife to live apart. He wrote altogether upon his West African adventures, the enormous number of 9 volumes! namely: *Wanderings in West Africa* (2 vols.), *Abeokuta and the Cameroons* (2 vols.), *A Mission to the King of Dahomé* (2 vols.), *Wit and Wisdom from West Africa* (1 vol.), *Two Trips to Gorilla Land and the Cataracts of the Congo* (2 vols.). Remorselessly condensed, these nine might, with artistry, have made a book worthy to live. But Burton's prolixity is his reader's despair. He was devoid of the faintest idea of proportion. Consequently at the present day his books are regarded as mere quarries. He dedicated his *Abeokuta* "To my best friend," my wife, with a Latin verse which has been rendered :

> "Oh, I could live with thee in the wild wood
> Where human foot hath never worn a way ;
> With thee, my city, and my solitude,
> Light of my night, sweet rest from cares by day."

In her own copy Mrs. Burton wrote close to the lines, "Thank you, sweet love!"[1]

Burton and his wife now set out for Lisbon, where they saw a bull-fight, because Burton said people "ought to see everything once," though this did not prevent them from going to several other bull-fights. Mrs. Burton was not at all afraid of the bulls, but when some cockroaches invaded her apartment she got on a chair and screamed, though even then they did not go away. More than that, numbers of other cockroaches came to see what was the matter; and they never left off coming. After "a delightful two months" at Lisbon, Burton set out for Brazil, while his wife returned to England "to pay and pack." She rejoined him some weeks later at Rio Janeiro, and they reached Santos on 9th October 1865. They found it a plashy, swampy place, prolific in mangroves and tree ferns, with here and there a cultivated patch. Settlers, however, became attached to it. Sandflies and mosquitoes abounded, and the former used to make Burton "come out all over lumps." Of the other vermin, including multitudinous snakes, and hairy spiders the size of toy terriers they took no particular notice. The amenities of the place were wonderful orchids, brilliantly coloured parrots and gigantic butterflies with great prismy wings. The Burtons kept a number of slaves, whom, however, they paid "as if they were free men," and Mrs. Burton erected a chapel for them—her oratory—where the Bishop "gave her leave to have mass and the sacraments." Her chief convert, and he wanted converting very badly, was an inhuman, pusillanimous coal-black dwarf, 35 years of age, called Chico,[2] who became her right-hand man. Just as she had made him to all appearance a good sound Catholic she caught him roasting alive her favourite cat before the kitchen fire. This was the result partly of innate diablery and partly of her

(1) The copy is in the Public Library, High Street, Kensington, where most of Burton's books are preserved.

(2) Spanish for "little one."

having spoilt him, but wherever she went Mrs. Burton managed to get a servant companion whom her lack of judgment made an intolerable burden to her. Chico was only the first of a series. Mrs. Burton also looked well after the temporal needs of the neighbourhood, but if she was always the Lady Bountiful, she was rarely the Lady Judicious.

52. Aubertin. Death of Steinhäuser, 27th July 1866.

The Burtons resided sometimes at Santos and sometimes at Sao Paulo, eight miles inland. These towns were just then being connected by railway; and one of the superintendents, Mr. John James Aubertin, who resided at Sao Paulo, became Burton's principal friend there. Aubertin was generally known as the "Father of Cotton," because during the days of the cotton famine, he had laboured indefatigably and with success to promote the cultivation of the shrub in those parts. Like Burton, Aubertin loved Camoens, and the two friends delighted to walk together in the butterfly-haunted forests and talk about the "beloved master," while each communicated to the other his intention of translating *The Lusiads* into English. Thirteen years, however, were to elapse before the appearance of Aubertin's translation[1] and Burton's did not see print till 1880. In 1866 Burton received a staggering blow in the loss of his old friend Dr. Steinhäuser, who died suddenly of heart disease, during a holiday in Switzerland, 27th July 1866. It was Steinhäuser, it will be remembered, with whom he had planned the translation of *The Arabian Nights*, a subject upon which they frequently corresponded.[2]

Wherever Burton was stationed he invariably interested himself in the local archæological and historical

(1) *The Lusiads*, 2 vols., 1878. Says Aubertin, "In this city (Sao Paulo) and in the same room in which I began to read *The Lusiads* in 1860, the last stanza of the last canto was finished on the night of 24th February 1877."

(2) Burton dedicated the 1st vol. of his *Arabian Nights* to Steinhäuser.

associations. Thus at Santos he explored the enormous kitchen middens of the aboriginal Indians; but the chief attraction was the site of a Portuguese fort, marked by a stone heap, where a gunner, one Hans Stade, was carried off by the cannibals and all but eaten. Burton used to visit the place by boat, and the narrative written by Hans Stade so fascinated him that he induced a Santos friend, Albert Tootal, to translate it into English. The translation was finished in 1869, and five years later Burton wrote for it an introduction and some valuable notes and sent it to press. Though Burton scarcely shines as an original writer, he had a keen eye for what was good in others, and he here showed for the first time that remarkable gift for annotating which stood him in such stead when he came to handle *The Arabian Nights*.

53. The Facetious Cannibals.

Hans Stade's story is so amusing that if we did not know it to be fact we should imagine it the work of some Portuguese W. S. Gilbert. Never were more grisly scenes or more captivating and facetious cannibals. When they told Stade that he was to be eaten, they added, in order to cheer him, that he was to be washed down with a really pleasant drink called kawi. The king's son then tied Stade's legs together in three places. "I was made," says the wretched man, "to hop with joined feet through the huts; at this they laughed and said 'Here comes our meat hopping along,'" Death seemed imminent. They did Stade, however, no injury beside shaving off his eyebrows, though the younger savages, when hungry, often looked wistfully at him and rubbed their midriffs. The other prisoners were, one by one, killed and eaten, but the cannibals took their meals in a way that showed indifferent breeding. Even the king had no table manners whatever, but walked about gnawing a meaty bone. He was good-natured, however, and offered a bit to Stade, who not only declined, but uttered some words of reproof. Though surprised, the king was not angry; he took another bite and observed critically, with his mouth full, "It tastes good!"

Life proceeded slowly, whether at Santos or Sao Paulo, almost the only excitement being the appearance of companies of friendly Indians. They used to walk in single file, and on passing Burton's house would throw out their arms as if the whole file were pulled by a string. Burton did not confine himself to Santos, however. He wandered all over maritime Brazil, and at Rio he lectured before the king [1] and was several times invited to be present at banquets and other splendid gatherings. On the occasion of one of these notable functions, which was to be followed by a dinner, one room of the palace was set apart for the ministers to wait in and another for the consuls. The Burtons were told not to go into the consular room, but into the ministers' room. When, however, they got to the door the officials refused to let them pass.

"This is the ministers' room," they said, "You cannot come here."

"Well, where am I to go?" enquired Burton.

Mrs. Burton stood fuming with indignation at the sight of the stream of nonentities who passed in without question, but Burton cried, "Wait a moment, my darling. I've come to see the Emperor, and see the Emperor I will."

So he sent in his card and a message.

"What!" cried the Emperor, "a man like Burton excluded. Bring him to me at once." So Burton and his wife were conducted to the Emperor and Empress, to whom Burton talked so interestingly, that they forgot all about the dinner. Meanwhile flunkeys kept moving in and out, anxiety on their faces—the princes, ambassadors and other folk were waiting, dinner was waiting; and the high functionaries and dinner were kept waiting for half an hour. "Well, I've had my revenge," said Burton to his wife when the interview was over. "Only

1) Dom Pedro, deposed 15th November 1889.

think of those starving brutes downstairs; but I'm sorry on your account I behaved as I did, for it will go against all your future 'at homes.'" At dinner the Emperor and the Empress were most attentive to the Burtons, and the Empress gave Mrs. Burton a beautiful diamond bracelet.[1]

Among Burton's admirers was a Rio gentleman named Cox, who had a mansion near the city. One day Mr. Cox arranged a grand dinner party and invited all his friends to meet the famous traveller. Burton arrived early, but presently disappeared. By and by the other guests streamed in, and after amusing themselves for a little while about the grounds they began to enquire for Burton. But no Burton was to be seen. At last someone happened to look up the highest tree in the compound and there was the guest of the day high among the branches squatting like a monkey. He had got up there, he said, to have a little peace, and to keep on with the book he was writing about Brazil. He came down, however, when the lunch bell rang, for though he grumbled at all other noises, he maintained that, somehow that sound always had a peculiar sweetness.

Wit and humour, wherever found, never failed to please Burton, and a remark which he heard in a Brazilian police court and uttered by the presiding magistrate, who was one of his friends, particularly tickled him:

"Who is this man?" demanded the magistrate, in reference to a dissipated-looking prisoner.

"Un Inglez bebado" (a drunken Englishman), replied the constable.

"A drunken Englishman," followed the magistrate, "What a pleonasm!"

A little later Burton and his wife went down a mine which ran three quarters of a mile into the earth. "The negret Chico," says Burton, "gave one glance at the deep,

[1] This anecdote differs considerably from Mrs. Burton's version, *Life*, i., 438. I give it, however, as told by Burton to his friends.

dark pit, wrung his hands and fled the Tophet, crying that nothing in the wide, wide world would make him enter such an Inferno. He had lately been taught that he is a responsible being, with an 'immortal soul,' and he was beginning to believe it in a rough, theoretical way: this certainly did not look like a place 'where the good niggers go.'" However, if Chico turned coward Burton and his wife did not hesitate. But they had moments of fearful suspense as they sank slowly down into the black abysm. The snap of a single link in the long chain would have meant instantaneous death; and a link had snapped but a few days previous, with fatal results. Arrived at the bottom they found themselves in a vast cave lighted with a few lamps—the walls black as night or reflecting slender rays from the polished watery surface. Distinctly Dantesque was the gulf between the huge mountain sides which threatened every moment to fall. One heard the click and thud of hammers, the wild chants of the borers, the slush of water. Beings like gnomes and kobolds glided hither and thither—half naked figures muffled up by the mist. Here dark bodies, gleaming with beaded heat drops, hung in what seemed frightful positions; "they swung like Leotard from place to place." Others swarmed up loose ropes like Troglodytes. It was a situation in which "thoughts were many and where words were few."

Burton and his wife were not sorry when they found themselves above ground again and in the sweet light of day.

54. Up the Sao Francisco.

The next event was a canoe journey which Burton made alone up the river Sao Francisco from its source to the falls of Paulo Affonso—and then on to the sea, a distance of 1500 miles—an astounding feat even for him. During these adventures a stanza in his own unpublished version of Camoens constantly cheered him:

> "Amid such scenes with danger fraught and pain
> Serving the fiery spirit more to flame,
> Who woos bright honour, he shall ever win
> A true nobility, a deathless fame

> Not they who love to lean, unjustly vain,
> Upon the ancestral trunk's departed claim;
> Nor they reclining on the gilded beds
> Where Moscow's zebeline downy softness spreads." (1)

Indeed he still continued, at all times of doubt and despondency, to turn to his beloved poet; and always found something to encourage.

The year before his arrival in Santos a terrible war had broken out between Brazil, Uruguay and Argentina on the one side and Paraguay on the other; 55. In Paraguay. August 15th to September 15th 1868. April 4th to April 18th 1869.
the Paraguayan dictator Lopez II. had been defeated in many battles and Paraguay so long, thanks to the Jesuits and Dr. Francia, a thriving country, was gradually being reduced to ruin. Tired of Santos, which was out of the world and led to nothing, Burton in July 1868 sent in his resignation. Mrs. Burton at once proceeded to England, but before following her, Burton at the request of the Foreign Office, travelled through various parts of South America in order to report the state of the war. He visited Paraguay twice, and after the second journey made his way across the continent to Arica in Peru, whence he took ship to London via the Straits of Magellan.(2) During part of the voyage he had as fellow traveller Arthur Orton, the Tichborne claimant. As both had spent their early boyhood at Elstree they could had they so wished have compared notes, but we may be sure Mr. Orton preserved on that subject a discreet silence. The war terminated in March 1870, after the death of Lopez II. at the battle of Aquidaban. Four-fifths of the population of Paraguay had perished by sword or famine.

(1) *Lusiads*, canto 6, stanza 95. Burton subsequently altered and spoilt it. The stanza as given will be found on the opening page of the Brazil book.

(2) He describes his experiences in his work *The Battlefields of Paraguay*.

CHAPTER XIV

OCTOBER 1869—16TH AUGUST 1871

"EMPEROR AND EMPRESS OF DAMASCUS."

BIBLIOGRAPHY:

32. Vikram and the Vampire.
33. Letters from the Battlefields of Paraguay. 1870.
34. Proverba Communia Syriaca. 1871.
35. The Jew. Written 1871, published 1898.

56. Archbishop Manning and the Odd Fish.

Mrs. Burton had carried with her to England several books written by her husband in Brazil, and upon her arrival she occupied herself first in arranging for their publication, and secondly in trying to form a company to work some Brazilian mines for which Burton had obtained a concession. The books were *The Highlands of Brazil* (2 vols. 1869), *The Lands of the Cazembé* (1873) and *Iracema, or Honey Lips*, a translation from the Brazilian (1886).

We hear no more of the mines, but she was able to send her husband "the excellent news of his appointment to the Consulate of Damascus." He heard of it first, however, not from her letter, but casually in a café at Lima, just as he was preparing to return home. On arriving in England almost his first business was to patent a pistol which he had invented especially for the use of travellers, and then he and Mrs. Burton gave themselves the pleasure of calling on old friends and going into society. To this date should, perhaps, be assigned the story[1] of Archbishop, afterwards Cardinal Manning, and

(1) Unpublished. Told me by Mrs. E. J. Burton. Manning was made a cardinal in 1875.

the Odd Fish. Burton had just presented to the Zoological Gardens a curious fish which lived out of water, and took but little nourishment. He had often presented different creatures to the Zoo, though nobody had ever even thanked him, but this gift created some commotion, and "Captain Burton's Odd Fish" became the talk of London.

In the midst of its popularity Burton one day found himself seated at a grand dinner next to his good friend the long, lean and abstemious Archbishop Manning. But much as Burton liked Manning, he could never bear to be near him at meal times. Manning always would eat little and talk much; so Burton, who was a magnificent trencherman, suffered serious inconvenience, and the present occasion proved no exception. It was in vain that Burton urged the Archbishop to mortify himself by eating his dinner. After a while Mrs. Burton, who sat on the other side of the Archbishop, remarked "Richard must take you to the Zoo and show you his famous fish." "I'll certainly go," said Manning, turning to Burton, "I am really curious to see it." "Then my Lord," followed Burton, "there will be a pair of odd fish. You know, you neither eat nor drink, and that's the peculiarity of the other fish."

As usual when in England, Burton spoke at several public meetings, and Mrs. Burton, of whose appearance he continued to be justifiably proud, generally accompanied him on the platform. Before speaking he always ate sparingly, saying "No" to almost everything. On one of such evenings he was the guest of Dr. Burton, and by chance, hot curry, his favourite dish, was placed on the table. "Now this is real wickedness, cousin," he exclaimed, "to have hot curry when I can't eat it." When dinner was nearly over somebody came in with a basket of damask roses. "Ask for two of them," whispered Burton to his wife. "She did, and appeared with them in her bosom on the platform, "And oh," added my informer, "how handsome she looked!"

Having visited Uriconium, the English Pompeii, the Burtons made for Vichy, where they met Mr. Swinburne, (Sir) Frederick Leighton and Mrs. Sartoris. His companions on this journey, as on so many others, were two books—one being the anodynous Camoens, the other a volume consisting of the Bible, Shakespeare and Euclid bound together, which looked, with its three large clasps, like a congested Church Service. Mrs. Burton then returned to England "to pay and pack," while Burton, "being ignorant" as they say in the *Nights*, "of what lurked for him in the secret purpose of God," proceeded to Damascus, with two bull-terriers, descendants, no doubt, of the Oxford beauty.

Mrs. Burton followed in December, with her entire fortune—a modest £300 in gold, and life promised to be all labdanum. Disliking the houses in

57. 3rd Consulate, Damascus.

Damascus itself, the Burtons took one in the suburb El Salahiyyah; and here for two years they lived among white domes and tapering minarets, palms and apricot trees. Midmost the court, with its orange and lemon trees, fell all day the cool waters of a fountain. The principal apartments were the reception room, furnished with rich Eastern webs, and a large dining room, while a terrace forming part of the upper storey served as "a pleasant housetop in the cool evenings." The garden, with its roses, jessamine, vines, citron, orange and lemon trees, extended to that ancient river, the jewel-blue Chrysorrhoa. There was excellent stabling, and Mrs. Burton kept horses, donkeys, a camel, turkeys, bull-terriers, street dogs, ducks, leopards, lambs, pigeons, goats, and, to use Burton's favourite expression, "other notions." They required much patient training, but the result was satisfactory, for when most of them had eaten one another they became a really harmonious family.

If Mrs. Burton went abroad to the bazaar or elsewhere she was accompanied by four Kawwasses in full dress of scarlet and gold, and on her reception day these gorgeous attendants kept guard. Her visitors sat on the divans cross-legged or not according to their nation, smoked, drank sherbet and coffee, and ate sweetmeats.

For Ra'shid Pasha, the Wali or Governor-General of Syria, both Burton and his wife conceived from the first a pronounced antipathy. He was fat and indolent, with pin-point eyes, wore furs, walked on his toes, purred and looked like "a well-fed cat." It did not, however, occur to them just then that he was to be their evil genius

"Call him Ra'shid, with the accent on the first syllable," Burton was always careful to say when speaking of this fiendish monster, "and do not confound him with (Haroun al) Rashi'd, accent on the second syllable—'the orthodox,' the 'treader in the right path.'"[1]

At an early date Burton formed a friendship with the Algerine hero and exile Abd el Kadir, a dark, kingly-looking man who always appeared in snow white and carried superbly-jewelled arms; while Mrs. Burton, who had a genius for associating herself with undesirable persons, took to her bosom the notorious and polyandrous Jane Digby el Mezrab.[2] This lady had been the wife first of Lord Ellenborough, who divorced her, secondly of Prince Schwartzenberg, and afterwards of about six other gentlemen. Finally, having used up Europe, she made her way to Syria, where she married a "dirty little black"[3] Bedawin shaykh. Mrs. Burton, with her innocent, impulsive, flamboyant mind, not only grappled Jane Digby with hoops of steel, but stigmatised all the charges against her as wilful and malicious. Burton, however, mistrusted the lady from the first. Says Mrs. Burton of her new friend, "She was a most beautiful woman, though sixty-one, tall, commanding, and queen-like.

58. Jane Digby el Mezrab.

(1) Mr. John Payne, however, proves to us that the old Rashi'd, though a lover of the arts, was also a sensual and bloodthirsty tyrant. See Terminal Essay to his *Arabian Nights*, vol. ix.

(2) She thus signed herself after her very last marriage.

(3) Mrs. Burton's words.

She was *grande dame jusqu' au bout des doigts*, as much as if she had just left the salons of London and Paris, refined in manner, nor did she ever utter a word you could wish unsaid. She spoke nine languages perfectly, and could read and write in them. She lived half the year in Damascus and half with her husband in his Bedawin tents, she like any other Bedawin woman, but honoured and respected as the queen of her tribe, wearing one blue garment, her beautiful hair in two long plaits down to the ground, milking the camels, serving her husband, preparing his food, sitting on the floor and washing his feet, giving him his coffee; and while he ate she stood and waited on him: and glorying in it. She looked splendid in Oriental dress. She was my most intimate friend, and she dictated to me the whole of her biography." (1) Both ladies were inveterate smokers, and they, Burton, and Abd el Kadir spent many evenings on the terrace of the house with their narghilehs. Burton and his wife never forgot these delightsome causeries. Swiftly, indeed, flew the happy hours when they

"Nighted and dayed in Damascus town." (2)

Burton had scarcely got settled at Damascus before he expressed his intention of visiting the historic Tadmor in the desert. It was an eight days' journey, and the position of the two wells on the way was kept a secret by Jane Digby's tribe, who levied blackmail on all visitors to the famous ruins. The charge was the monstrous one of £250; but Burton—at all times a sworn foe to cupidity—resolved to go without paying. Says Mrs. Burton, "Jane Digby was in a very anxious state when she heard this announcement, as she knew it was a death blow to a great source of revenue to the tribe. . . She did all she could to dissuade us, she wept over our loss, and she told us that we should never come back." Finally the subtle lady dried her crocodile eyes and offered her "dear friends" the escort of one of her Bedawin,

59. To Tadmor.

(1) *Life* i., p. 486.

(2) *Arabian Nights.* Lib. Ed, i., 215.

that they might steer clear of the raiders and be conducted more quickly to water, "if it existed." Burton motioned to his wife to accept the escort, and Jane left the house with ill-concealed satisfaction. The Bedawi [1] in due time arrived, but not before he had been secretly instructed by Jane to lead the Burtons into ambush whence they could be pounced upon by the tribe and kept prisoners till ransomed. That, however, was no more than Burton had anticipated; consequently as soon as the expedition was well on the road he deprived the Bedawi of his mare and accoutrements, and retained both as hostages until Damascus should be reached again. Appropriately enough this occurred on April the First.[2] Success rewarded his acuteness, for naturally the wells were found, and the travellers having watered their camels finished the journey with comfort. Says Mrs. Burton, "I shall never forget the imposing sight of Tadmor. There is nothing so deceiving as distance in the desert. . . . A distant ruin stands out of the sea of sand, the atmosphere is so clear that you think you will reach it in half an hour; you ride all day and you never seem to get any nearer to it." Arrived at Tadmor they found it to consist of a few orchards, the imposing ruins, and a number of wretched huts "plastered like wasps' nests within them." Of the chief ruin, the Temple of the Sun, one hundred columns were still standing and Burton, who set his men to make excavations, found some statues, including one of Zenobia. The party reached Damascus again after an absence of about a month. The Bedawi's mare was returned; and Jane Digby had the pleasure of re-union with her dear Mrs. Burton, whom she kissed effusively.

Both Burton and his wife mingled freely with the people of Damascus, and Burton, who was constantly storing up knowledge against his great edition of *The*

(1) Burton generally writes Bedawi and Bedawin. Bedawin (Bedouin) is the plural form of Bedawi. *Pilgrimage to Meccah,* vol. ii, p. 80.

(2) 1870. Three months after Mrs. Burton's arrival.

Arabian Nights, often frequented the Arabic library.[1] Their favourite walk was to the top of an adjacent eminence, whence they could look down on Damascus, which lay in the light of the setting sun, "like a pearl." Then there were excursions to distant villages of traditionary interest, including Jobar, where Elijah is reputed to have hidden, and to have anointed Hazael.[2] "The Bird," indeed, as ever, was continually on the wing, nor was Mrs. Burton less active. She visited, for example, several of the harems in the city, including that of Abd el Kadir. "He had five wives," she says, "one of them was very pretty. I asked them how they could bear to live together and pet each other's children. I told them that in England, if a woman thought her husband had another wife or mistress, she would be ready to kill her. They all laughed heartily at me, and seemed to think it a great joke."[3] She also took part in various social and religious functions, and was present more than once at a circumcision—at which, she tells us, the victim, as Westerns must regard him, was always seated on richest tapestry resembling a bride throne, while his cries were drowned by the crash of cymbals. Burton's note books, indeed, owed no mean debt to her zealous co-operation.

60. Palmer and Drake. 11th July 1870.

The Burtons spent their summer in a diminutive Christian village called B'ludan, on the Anti-Lebanon, at the head of the Vale of Zebedani, Burton having chosen it as his sanitarium. A beautiful stream with waterfalls bubbled through their gardens, which commanded magnificent views of the Lebanon country. As at Santos, Mrs. Burton continued her rôle of Lady Bountiful, and she spent many hours making up powders and pills. Although in reality

(1) It contained, among other treasures, a Greek manuscript of the Bible with the Epistle of Barnabas and a portion of the Shepherd of Hermas.

(2) 1 *Kings*, xix., 15; 2 *Kings*, viii., 15.

(3) *The Romance of Isabel Lady Burton*, ii., 386.

nobody was one jot the better or the worse for taking them, the rumour circulated that they were invariably fatal. Consequently her reputation as a doctor spread far and wide. One evening a peasant woman who was dying sent a piteous request for aid, and Mrs. Burton, who hurried to the spot, satisfied the poor soul by the administration of some useless but harmless dose. Next morning the woman's son appeared. He thanked Mrs. Burton warmly for her attentions, said it was his duty to report that his mother was dead, and begged for a little more of the efficacious white powder, as he had a bedridden grandmother of whom he was also anxious to be relieved.

One piping hot morning [1] when walking in his garden Burton noticed a gipsy tent outside, and on approaching it found two sun-burnt Englishmen, a powerful, amiable-looking giant, and a smaller man with a long beard and silky hair. The giant turned out to be Charles Tyrwhitt Drake and the medium-sized man Edward Henry Palmer, both of whom were engaged in survey work. Drake, aged 24, was the draughtsman and naturalist; Palmer, [2] just upon 30, but already one of the first linguists of the day, the archæologist. Palmer, like Burton, had leanings towards occultism: crystal gazing, philosopher's stone hunting. After making a mess with chemicals, he would gaze intently at it, and say excitedly: "I wonder what will happen"— an expression that was always expected of him on such and all other exciting occasions. A quadruple friendship ensued, and the Burtons, Drake and Palmer made several archæological expeditions together. To Palmer's poetical eyes all the Lebanon region was enchanted ground. Here the lovely Shulamite of the lovelier Scripture lyric fed her flocks by the shepherd's tents. Hither came Solomon, first

(1) 11th July 1870.

(2) E. H. Palmer (1840—1882). In 1871 he was appointed Lord Almoner's Professor of Arabic at Cambridge. He was murdered at Wady Sudr, 11th August 1882. See Chapter xxiii.

disguised as a shepherd, to win her love, and afterwards in his royal litter perfumed with myrrh and frankincense to take her to his Cedar House. This, too, was the country of Adonis. In Lebanon the wild boar slew him, and yonder, flowing towards "holy Byblus," were "the sacred waters where the women of the ancient mysteries came to mingle their tears."(1) Of this primitive and picturesque but wanton worship they were reminded frequently both by relic and place name. To Palmer, viewing them in the light of the past, the Cedars of Lebanon were a poem, but to Burton—a curious mixture of the romantic and the prosaic—with his invariable habit of underrating famous objects, they were "a wretched collection of scraggy Christmas trees." "I thought," said Burton, "when I came here that Syria and Palestine would be so worn out that my occupation as an explorer was clean gone." He found, however, that such was not the case—all previous travellers having kept to the beaten tracks; Jaydur, for example, the classical Ituræa, was represented on the maps by "a virgin white patch." Burton found it teeming with interest. There was hardly a mile without a ruin—broken pillars, inscribed slabs, monoliths, tombs. A little later he travelled as far northward as Hamah(2) in order to copy the uncouth characters on the famous stones, and Drake discovered an altar adorned with figures of Astarte and Baal.(3) Everywhere throughout Palestine he had to deplore the absence of trees. "Oh that Brigham Young were here!" he used to say, "to plant a million. The sky would then no longer be brass, or the face of the country a quarry." Thanks to his researches, Burton has made his name historical in the Holy Land, for his book *Unexplored Syria*—written though it be in a distressingly slipshod style—throws, from almost every

(1) Renan. See, too, *Paradise Lost*, Bk. 1. *Isaiah* (xvii., 10) alludes to the portable "Adonis Gardens" which the women used to carry to the bier of the god.

(2) The Hamath of Scripture. 2 *Sam.*, viii., 9; *Amos*, vi., 2.

(3) See illustrations in *Unexplored Syria*, by Burton and Drake.

page, interesting light on the Bible. "Study of the Holy Land," he said, "has the force of a fifth Gospel, not only because it completes and harmonises, but also because it makes intelligible the other four. Oh, when shall we have a reasonable version of Hebrew Holy Writ which will retain the original names of words either untranslatable or to be translated only by guess work! [1] One of their adventures—with a shaykh named Salámeh—reads like a tale out of *The Arabian Nights*. Having led them by devious paths into an uninhabited wild, Salámeh announced that, unless they made it worth his while to do otherwise, he intended to leave them there to perish, and it took twenty-five pounds to satisfy the rogue's cupidity. Palmer, however, was of opinion that an offence of this kind ought by no means to be passed over, so on reaching Jerusalem he complained to the Turkish governor and asked that the man might receive punishment. "I know the man," said the Pasha, "he is a scoundrel, and you shall see an example of the strength and equity of the Sultan's rule;" and of course, Palmer, in his perpetual phrase, wondered what would happen. After their return to Damascus the three friends had occasion to call on Rashid Pasha. "Do you think," said the Wali, with his twitching moustache and curious, sleek, unctuous smile, "do you think you would know your friend again?" He then clapped his hands and a soldier brought in a sack containing four human heads, one of which had belonged to the unfortunate Salámeh. "Are you satisfied?" enquired the Wali. [2]

61. Khamoor.

Having been separated from "that little beast of a Brazilian"—the cat-torturing Chico—Mrs. Burton felt that she must have another confidential servant companion. Male dwarfs being so unsatisfactory she now decided to try a full-sized human being, and of the other sex. At Miss Ellen Wilson's

(1) *The Land of Midian Revisited*, ii., 73.

(2) *Life of Edward H. Palmer*, p. 109.

Protestant Mission in Anti-Lebanon she saw just her ideal—a lissom, good-looking Syrian maid, named Khamoor, or "The Moon." Chico the Second (or shall we say Chica [1] the First.) had black plaits of hair confined by a coloured handkerchief, large, dark, reflulgent eyes pouting lips, white teeth, of which she was very proud, "a temperament which was all sunshine and lightning in ten minutes,' and a habit of discharging, quite un-unexpectedly, a "volley of fearful oaths." She was seventeen—"just the time of life when a girl requires careful guiding." So Mrs. Burton, or "Ya Sitti," as Khamoor called her, promptly set about this careful guiding—that is to say she fussed and petted Khamoor till the girl lost all knowledge of her place and became an intolerable burden. Under Mrs. Burton's direction she learnt to wear stays [2] though this took a good deal of learning; and also to slap men's faces and scream when they tried to kiss her. By dint of practice she in time managed this also to perfection. Indeed, she gave up, one by one, all her heathenish ways, except swearing, and so became a well-conducted young lady, and almost English. Mrs. Burton was nothing if not a woman with a mission, and henceforward two cardinal ideas swayed her namely, first to inveigle the heathen into stays, and secondly, to induce them to turn Catholics. Her efforts at conversion were more or less successful, but the other propaganda had, to her real sorrow, only barren results.

In March 1871, Charles Tyrwhitt Drake, who had spent some months in England, arrived again in Damascus, and the Burtons begged him to be their permanent guest. Henceforth Mrs. Burton, Burton and Drake were inseparable companions, and they explored together "almost every known part of Syria." Mrs. Burton used to take charge of the camp "and visited the harems to note things hidden from mankind," Drake sketched and collected botanical and geological specimens, while

(1) Chica is the feminine of Chico (Spanish).

(2) Mrs. Burton's expression.

KHAMOOR, LADY BURTON'S SYRIAN MAID.

Hutchinson & Co., Paternoster Row

Burton's studies were mainly anthropological and archæological. They first proceeded to Jerusalem, where they spent Holy Week, and after visiting Hebron, the Dead Sea, and other historical spots, they returned by way of Nazareth. But here they met with trouble. Early in his consulate, it seems, Burton had protested against some arbitrary proceedings on the part of the Greek Bishop of Nazareth, and thus made enemies among the Greeks. Unhappily, when the travellers appeared this ill-feeling led a posse of Nazarenes to make an attack on Burton's servants; and Burton and Drake, who ran half dressed out of their tents to see what was the matter, were received with a shower of stones, and cries of "Kill them!" Burton stood perfectly calm, though the stones hit him right and left, and Drake also displayed cool bravery. Mrs. Burton then hastened up with "two six shot revolvers," but Burton, having waved her back—snatched a pistol from the belt of one of his servants and fired it into the air, with the object of summoning his armed companions, whereupon the Greeks, though they numbered at least a hundred and fifty, promptly took to their heels. Out of this occurrence, which Burton would have passed over, his enemies, as we shall see, subsequently made considerable capital. The party then proceeded to the Sea of Galilee, whence they galloped across "their own desert" home. During these travels Burton and Drake made some valuable discoveries and saw many extraordinary peoples, though none more extraordinary than the lazy and filthy Troglodytes of the Hauran,[1] who shared the pre-historic caves with their cows and sheep, and fed on mallows just as their forefathers are represented as having done in the vivid thirtieth chapter of Job,[2] and in the pages of Agatharchides.[3]

(1) District east of the Sea of Galilee.

(2) *Job*, chapter xxx. "But now they that are younger than I have me in derision . . . who cut up mallows by the bushes and juniper roots for their meat."

(3) Greek Geographer. 250 B.C.

Mrs. Burton now heard news that fired her with joy. A sect of the Mohammedans called Shazlis used to assemble in the house of one of their number for Moslem prayer, reading and discussion.

62. The Shazlis.

One day they became conscious of a mysterious presence among them. They heard and saw things incommunicably strange, and a sacred rapture diffused itself among them. Their religion had long ceased to give them satisfaction, and they looked anxiously round in search of a better. One night when they were overcome by sleep there appeared to each a venerable man with a long white beard, who said sweetly, "Let those who want the truth follow me," and forthwith they resolved to search the earth until they found the original of the vision. But they had not to go far. One of them chancing to enter a monastery in Damascus noticed a Spanish priest named Fray Emanuel Förner. Hurrying back to his comrades he cried "I have seen the oldster of the dreams." On being earnestly requested to give direction, Förner became troubled, and with a view to obtaining advice, hurried to Burton. Both Burton and his wife listened to the tale with breathless interest. Mrs. Burton naturally wanted to sweep the whole sect straightway into the Roman Church, and it is said that she offered to be sponsor herself to 2,000 of them. In any circumstances, she distributed large numbers of crucifixes and rosaries. Burton, who regarded nine-tenths of the doctrines of her church as a tangle of error, was nevertheless much struck with the story. He had long been seeking for a perfect religion, and he wondered whether these people had not found it. Here in this city of Damascus, where Our Lord had appeared to St. Paul, a similar apparition had again been seen—this time by a company of earnest seekers after truth. He determined to investigate. So disguised as a Shazli, he attended their meetings and listened while Förner imparted the principal dogmas of the Catholic faith. His common sense soon told him that the so-called miraculous sights were merely hallucinations, the outcome of heated and hysterical imagination. He sympathised with the Shazlis in that like himself they were seekers after truth, and there, as far as he was concerned, the matter would have ended had the scenes been in any other country. But in Syria religious freedom was unknown, and the cruel Wali Rashid Pasha was only too delighted to have an opportunity to use his power. He crushed where he could not controvert. Twelve of the leading Shazlis—the martyrs, as they were called—were seized and imprisoned. Förner died suddenly; as some think, by poison. This threw Burton, who hated oppression in all its forms, into a towering rage, and he straightway flung the whole of his weight into the cause of the Shazlis. Persecution gave them holiness. He wrote to Lord Granville that there were at least twenty-five thousand Christians longing secretly for baptism, and he suggested methods by which they might be protected. He also recommended the Government to press upon the Porte many other reforms. Both Burton and his wife henceforward openly protected the Shazlis, and in fact made themselves, to use the words of a member of the English Government, "Emperor and Empress of Damascus."

That Rashid Pasha and his crawling myrmidons were rascals of the first water and that the Shazlis were infamously treated is very evident. It is also clear that Burton was more just than diplomatic. We cannot, however, agree with those who lay all the blame on Mrs. Burton. We may not sympathise with her religious views, but, of course, she had the same right to endeavour to extend her own church as the Protestants at Beyrout, who periodically sent enthusiastic agents to Damascus, had to extend theirs.

The Shazli trouble alone, however, would not have shaken seriously Burton's position; and whatever others may have thought, it is certain Burton himself never at any time in his life considered that in this matter any particular blame attached to his wife. But unfortunately the Shazli trouble was only one of a series. Besides embroiling himself with the truculent Rashid Pasha and his underlings, Burton contrived to give offence to four other bodies of men. In June, 1870, Mr. Mentor Mott, the kind and charitable[1] superintendent of the British Syrian School at Beyrout, went to Damascus to proselytize, and acted, in Burton's opinion, with some indiscretion. Deeming Damascus just then to be not in a temper for proselytising, Burton reprimanded him, and thus offended the Protestant missionaries and Mr. Jackson Eldridge, the Consul-General at Beyrout. In Burton's opinion, but for Mrs. Mott the storm would have gradually subsided. That lady, however, took the matter more to heart than her husband, and was henceforth Burton's implacable enemy. Then arose a difficulty with the Druzes, who had ill-treated some English missionaries. As they were Turkish subjects the person to act was Rashid Pasha, but Burton and he being at daggers drawn, Burton attempted to fine the Druzes himself. He was reminded, however, that his power was limitary, and that he would not be allowed to exceed it. To the trouble with the Greeks we have already referred. But his chief enemies were the Jews, or rather the Jewish money-lenders, who used to go to the distressed villages, offer money, keep all the papers, and allow their victims nothing to show. Interest had to be paid over and over again. Compound interest was added, and when payment was impossible the defaulters were cast into prison. Burton's predecessor had been content to let matters alone, but Burton's blood boiled when he thought of these enormities. Still, when the money-lenders came to him and stated their case, he made for a time an honest attempt to double; but ultimately his indignation got the better of his diplomacy, and with an oath that made the windows rattle, he roared, "Do you think I am going to be bum-bailiff to a parcel of blood-suckers!" And yet these gentlemen had sometimes, in their moderation, charged as little as sixty per cent. Henceforward Burton looked

(1) Burton's words.

evil upon the whole Jewish race, and resolved to write a book embodying his researches respecting them and his Anti-Semite opinions. For the purpose of it he made minute enquiries concerning the death of one Padre Tommaso, whom the Jews were suspected of having murdered in 1840. These enquiries naturally gave his foes further umbrage, and they in return angrily discharged their venom at him. In his book *The Jew*, published after his death,[1] he lashes the whole people. He seems in its pages to be constantly running up and down with a whip and saying: "I'll teach you to be 'an Ebrew Jew,' I will." His credulity and prejudice are beyond belief. He accepts every malicious and rancorous tale told against the Jews, and records as historical facts even such problematical stories as the murder of Hugh of Lincoln. Thus he managed to exasperate representatives of almost every class. But perhaps it was his championship of the Shazlis that made the most mischief. Says Lady Burton, "It broke his career, it shattered his life, it embittered him towards religion."

Complaints and garbled stories reached London from all sides, and Burton was communicated with. He defended himself manfully, and showed that in every question he had been on the side of righteousness and equity, that he had simply fought systematically against cruelty, oppression and nefariousness. He could not and would not temporize. An idea of the corruption prevalent at Damascus may be gathered from the fact that on one occasion £10,000 was promised him if he would "give an opinion which would have swayed a public transaction." Says Lady Burton, "My husband let the man finish, and then he said, 'If you were a gentleman of my own standing, and an Englishman, I would just pitch you out of the window; but as you are not, you may pick up your £10,000 and walk down the stairs.'"[2]

(1) Published in 1898.

(2) *Life*, i., 572.

Accusations, many of them composed of the bluest gall; and manly letters of defence from Burton now flew almost daily from Damascus to England. The Wali, the Jews and others all had their various grievances. As it happened, the British Government wanted, just then, above all things, peace and quiet. If Burton could have managed to jog along in almost any way with the Wali, the Druzes, the Greeks, the Jews, and the other factors in Syria, there would have been no trouble. As to whether Burton was right or wrong in these disputes, the Government seems not to have cared a straw or to have given a moment's thought. Here, they said, is a man who somehow has managed to stir up a wasp's nest, and who may embroil us with Turkey. This condition of affairs must cease. Presently came the crash. On August 16th just as Burton and Tyrwhitt Drake were setting out for a ride at B'ludan, a messenger appeared and handed Burton a note. He was superseded. The blow was a terrible one, and for a moment he was completely unmanned. He hastened to Damascus in the forlorn hope that there was a mistake. But it was quite true, the consulship had been given to another

63. The Recall. 16th August 1871.

To his wife he sent the message, "I am superseded. Pay, pack, and follow at convenience." Then he started for Beirut, where she joined him. "After all my service," wrote Burton in his journal, "ignominiously dismissed at fifty years of age." One cry only kept springing from Mrs. Burton's lips, "Oh, Rashid Pasha! Oh Rashid Pasha!"

At Damascus Burton had certainly proved himself a man of incorruptible integrity. Even his enemies acknowledged his probity. But this availed nothing. Only two years had elapsed since he had landed in Syria, flushed with high premonitions; now he retired a broken man, shipwrecked in hope and fortune. When he looked back on his beloved Damascus—"O, Damascus, pearl of the East"—it was with the emotion evinced by the last of the Moors bidding adieu to Granada, and it only added to his exasperation when he imagined the exultation of the hated Jews, and the sardonic grin on the sly, puffy, sleek face of Rashid Pasha.

Just before Mrs. Burton left B'ludan an incident occurred which brings her character into high relief. A dying Arab boy was brought to her to be treated for rheumatic fever. She says, "I saw that death was near. . . . 'Would you like to see Allah?' I said, taking hold of his cold hand. . . . I parted his thick, matted hair, and kneeling, I baptised him from the flask of water I always carried at my side. 'What is that?' asked his grandmother after a minute's silence. 'It is a blessing,' I answered, 'and may do him good!'"[1] The scene has certain points in common with that enacted many years after in Burton's death chamber. Having finished all all her "sad preparations at B'ludan," Mrs. Burton "bade adieu to the Anti-Lebanon with a heavy heart, and for the last time, choking with emotion, rode down the mountain and through the Plain of Zebedani, with a very large train of followers."—"I had a sorrowful ride," says she, "into Damascus. Just outside the city gates I met the Wali, driving in state, with all his suite. He looked radiant, and saluted me with much *empressement*. I did not return his salute."[2]

It is satisfactory to know that Rashid Pasha's triumph was short-lived. Within a month of Burton's departure he was recalled by the Porte and disgraced. Not only so but every measure which Burton had recommended during his consulship was ordered to be carried out, and "The reform was so thorough and complete, that Her Majesty's Ambassador at Constantinople was directed officially to compliment the Porte upon its newly initiated line of progress." But nobody thanked, or even thought of Burton. On the occasion of his departure Burton received shoals of letters from prominent men of "every

(1) *The Romance of Isabel Lady Burton*, ii., 504.

(2) *The Romance of Isabel Lady Burton*, ii., 505.

creed, race and tongue,'' manifesting sorrow and wishing him God-speed. Delightful, indeed, was the prologue of that from Abd El Kadir: "Allah," it ran, "favour the days of your far-famed learning, and prosper the excellence of your writing. O wader of the seas of knowledge, O cistern of learning of our globe, exalted above his age, whose exaltation is above the mountains of increase and our rising place, opener by his books of night and day, traveller by ship and foot and horse, one whom none can equal in travel." The letter itself was couched in a few simple, heartfelt words, and terminated with "It is our personal friendship to you which dictates this letter." "You have departed, "wrote a Druze shaykh, "leaving us the sweet perfume of charity and noble conduct in befriending the poor and supporting the weak and oppressed, and your name is large on account of what God has put into your nature.'

Some of the authorities at home gave out that one of the reasons for Burton's recall was that his life was in danger from the bullets of his enemies, but Burton commented drily: "I have been shot at, at different times, by at least forty men who fortunately could not shoot straight. Once more would not have mattered much."

CHAPTER XV

16TH AUGUST 1871—4TH JUNE 1872

"THE BLACKNESS OF DARKNESS"

Arrived in England Burton went straight to his sister's at Norwood. His dejection was abysmal. Says Miss Stisted. "Strong, brave man though he was, the shock of his sudden recall told upon him cruelly. Not even during his last years, when his health had all but given way, was he so depressed. Sleep being impossible, he used to sit up, sometimes alone, sometimes with Sir H. Stisted, until the small hours of the morning, smoking incessantly. Tragedy was dashed with comedy; one night a terrible uproar arose. The dining-room windows had been left open, the candles alight, and the pug asleep under the table forgotten. A policeman, seeing the windows unclosed, knocked incessantly at the street door, the pug awoke and barked himself hoarse, and everyone clattered out of his or her bedroom to ascertain the cause of the disturbance. My uncle had quite forgotten that in quiet English households servants retire to rest before 3 a.m."(1) Subsequently Lady Stisted and her daughters resided at Folkestone, and thenceforth they were "the Folky folk." Burton also took an early opportunity to visit his brother, and tried to lead him into conversation; but nothing could break that Telamonian silence.

64. With Sir H. Stisted at Norwood. August 1871.

(1) *Temple Bar*, vol. xcii., p. 339.

Mrs. Burton, who had returned to Damascus "to pay and pack," now arrived in England, bringing with her very imprudently her Syrian maid Khamoor.

65. Reduced to £15.

The £16,000 left by Burton's father, the £300 Mrs. Burton took out with her, and the Damascus £1,200 a year, all had been spent. Indeed, Mrs. Burton possessed no more than the few pounds she carried about her person. In these circumstances prudence would have suggested leaving such a cipher as Khamoor in Syria, but that seems not to have occurred to her. It is probable, however, that the spendthrift was not she but her husband, for when she came to be a widow she not only proved herself an astute business woman, but accumulated wealth. On reaching London she found Burton "in one room in a very small hotel." His pride had not allowed him to make any defence of himself; and it was at this juncture that Mrs. Burton showed her grit. She went to work with all her soul, and for three months she bombarded with letters both the Foreign Office and outside men of influence. She was not discreet, but her pertinacity is beyond praise. Upon trying to learn the real reason of his recall, she was told only a portion of the truth. Commenting on one of the charges, namely that Burton "was influenced by his Catholic wife against the Jews," she said, "I am proud to say that I have never in my life tried to influence my husband to do anything wrong, and I am prouder still to say that if I had tried I should not have succeeded."

For ten months the Burtons had to endure "great poverty and official neglect," during which they were reduced to their last £15. Having been invited by Mrs. Burton's uncle, Lord Gerard, to Garswood,[1] they went thither by train. Says Mrs. Burton, "We were alone in a railway compartment, when one of the fifteen sovereigns rolled out of my purse, and slid between the boards of the carriage and the door, reducing us to £14. I sat on the floor and cried, and he sat by me with his arm round my

(1) Near St Helens, Lancs.

waist trying to comfort me."(1) The poet, as Keats tells us, "pours out a balm upon the world," and in this, his darkest hour, Burton found relief, as he had so often found it, in the pages of his beloved Camoens. Gradually his spirits revived, and he began to revolve new schemes. Indeed, he was never the man to sit long in gloom or to wait listessly for the movement of fortune's wheel. He preferred to seize it and turn it to his purpose.

66. An Orgie at Lady Alford's. 2nd November 1871.

If the Burtons lacked money, on the other hand they had wealthy relations with whom they were able to stay just as long as they pleased; and, despite their thorny cares, they threw themselves heartily into the vortex of society. Among their friends was Lady Marion Alford, a woman of taste, talent and culture. The first authority of the day on art needlework, she used to expound her ideas on the looms of the world from those of Circe to those of Mrs. Wheeler of New York. At one of Lady Alford's parties in her house at Princes Gate, October 1871, the Prince of Wales and the Duke of Edinburgh being present, Burton appeared dressed as a Syrian shaykh, and Mrs. Burton as a Moslem lady of Damascus. Burton was supposed not to understand English, and Mrs. Burton gave out that she had brought him over to introduce him to English society. She thus describes the occurrence in an unpublished letter to Miss Stisted.(2)

"Our orgie was great fun. The Bird and I wore Arab dresses. I went in the dress of an Arab lady of Damascus, but *as myself*, accompanied by Khamoor in her village dress and introducing Hadji Abdullah, a Moslem shaykh of Damascus. We then spoke only Arabic to each other, and the Bird broken French to the company present. We were twenty-eight at supper. The Prince of Wales and the Duke of Edinburgh were there. We let them into the joke, and they much enjoyed it, but all the rest

(1) *Life of Sir Richard Burton*, by Lady Burton, i., 591.

(2) 2nd November 1871.

were quite taken in half the evening. Even Lord Lyons and many of our old friends. The house was perfect and the fountain part [1] quite like Damascus. After supper we made Turkish coffee and narghilhis, and Khamoor handed them to the Princes on her knees, the tray on her head in Eastern fashion. They were delighted and spoke to her very kindly. They talked for long to Richard, and afterwards to me, and asked when we were going back to Syria *before Lord Granville's brother*." This letter, like most of Mrs. Burton's letters to Miss Stisted, is signed "Z," short for "Zoo."

In February (1872) Mrs. Burton's mother, who had for years been paralysed, grew rapidly worse. Says Mrs. Burton, writing to Miss Stisted (29th February), "My time is divided between her and Richard's concerns. She did rally a little and I took advantage of it to go out to one dinner and to the Thanksgiving Day [2] which we saw to perfection, and enjoyed enormously; and last night to a very large gathering at Lady Margaret Beaumont's. . . Everybody was there and it gave me an opportunity of saying 'How d'ye do?' to the world after my return from Syria. . . I am working tooth and nail at the Bird's [3] case, and have got our ambassador (Elliott) to see me at twelve next Saturday." At this time everyone was talking about Livingstone, the story of the meeting of him and Stanley being still fresh in men's minds. It was thought that another expedition ought to be sent out with Burton to lead, and a grand luncheon was got up for the express purpose of bringing Burton and a certain great personage together. When the soup was being served, the great personage, turning to Burton, said: "You are the man to go out to Livingstone. Come, consent, and I will contribute £500 to the expedition."

Mrs. Burton, who sat next to her husband, looked up

(1) The fountain was sculptured by Miss Hosmer.

(2) 27th February 1872. Celebration of the Prince of Wales's recovery from a six weeks' attack of typhoid fever.

(3) Her husband's case.

with beaming eyes, and her heart beat with joy. The object of the luncheon had been achieved, and Fortune was again bestowing her smiles; but as ill luck would have it, Burton happened just then to be in one of his contrary moods. He went on spooning up his soup, and, without troubling to turn his head, said, "I'll save your Royal Highness that expense."

Poor Mrs. Burton nearly fainted. The Livingstone expedition was subsequently undertaken by Cameron.

67. The Tichborne Trial.

Another event of this period was the Tichborne trial, but though Burton was subpœnaed by the claimant, his evidence really assisted the other side.

"I understand," began his interlocutor, "that you are the Central African traveller."

"I have been to Africa," modestly replied Burton.

"Weren't you badly wounded?"[1]

"Yes, in the back, running away."

His identity being established, Burton gave his evidence without further word fence. "When I went out to Brazil," he said, "I took a present from Lady Tichborne for her son, but being unable to find him,[2] I sent the present back. When returning from America, I met the claimant, and I recognise him simply as the man I met. That is all." Burton, like others, always took it for granted that the claimant obtained most of his information respecting the Tichbornes from Bogle, the black man, who had been in the service of the family.

68. Khamoor at the Theatre.

In some unpublished letters of Mrs. Burton, written about this time, we get additional references to Khamoor, and several of them are amusing. Says Mrs. Burton in one of them,[3] "Khamoor was charming at the theatre.

(1) Of course, this was an unnecessary question, for there was no mistaking the great scar on Burton's cheek; and Burton's name was a household word.

(2) February 1854. Sir Roger had sailed from Valparaiso to Rio Janeiro. He left Rio in the "Bella," which was lost at sea.

(3) Undated.

I cried at something touching, and she, not knowing why, flung herself upon my neck and howled. She nearly died with joy on seeing the clown, and said, 'Oh, isn't this delightful. What a lovely life! ' She was awfully shocked at the women dancing with 'naked legs,' and at all the rustic swains and girls embracing each other."

In January 1872, the Burtons were at Knowsley,(1) the Earl of Derby's, whence Mrs. Burton wrote an affectionate letter to Miss Stisted. She says,(2) "I hope you are taking care of yourself. Good people are scarce, and I don't want to lose my little pet." Later, Burton visited Lady Stisted at Edinburgh, and about that time met a Mr. Lock, who was in need of a trusty emissary to report on some sulphur mines in Iceland, for which he had a concession. The two came to terms, and it was decided that Burton should start in May. He spent the intervening time at Lord Gerard's,(3) and thence Mrs. Burton wrote to Miss Stisted(4) saying why she did not accompany Burton in his visit to his relatives. She says, "I hope you all understand that no animosity keeps me from Edinburgh. I should have been quite pleased to go if Richard had been willing, but I think he still fancies that Maria (Lady Stisted) would rather not see me, and I am quite for each one doing as he or she likes. . . . The Bird sends his fond love and a chirrup."

(1) Knowsley is close to Garswood, Lord Gerard's seat.

(2) Letter, 4th January 1872.

(3) Garswood, Newton-le-Willows, Lancashire.

(4) Unpublished letter.

CHAPTER XVI

4TH JUNE 1872—24TH OCTOBER 1872

IN ICELAND

BIBLIOGRAPHY:

36. Zanzibar: City, Island and Coast. 2 vols., 1872.
37. Unexplored Syria. 2 vols., 1872.
38. On Human Remains, etc., from Iceland, 1872.

69. In Edinburgh Again, 4th June 1872.

In May, Burton was back again in Edinburgh, preparing for the Iceland journey. He took many walks down Princes Street and up Arthur's Seat with Lady Stisted and his nieces, and "he was flattered," says Miss Stisted, "by the kindness and hospitality with which he was received. The 93rd Highlanders, stationed at the Castle, entertained in genuine Highland fashion; and at our house he met most of the leading Scotch families who happened to be lingering in the northern capital." Lord Airlie, the High Commissioner, held brilliant receptions at Holyrood. There were gay scenes—women in their smartest gowns, men wearing their medals and ribands. General Sir H. Stisted was there in his red collar and cross and star of the Bath. Burton "looked almost conspicuous in unadorned simplicity." On 4th June [1] Burton left for Iceland. The parting from his friends was, as usual, very hard. Says Miss Stisted, "His hands turned cold, his eyes filled with tears." Sir W. H. Stisted accompanied him to Granton, whence, with new hopes and aspirations, he set sail. Spectacularly, Iceland—Ultima Thule—as he calls it—was a disappointment to him. "The giddy, rapid rivers," were narrow

(1) *The True Life*, p. 336.

brooks, Hecla seemed but "half the height of Hermon," the Great Geyser was invisible until you were almost on the top of it. Its voice of thunder was a mere hiccough. Burton, the precise antithesis of old Sir John de Mandeville, was perhaps the only traveller who never told "travellers' tales." Indeed, he looked upon Sir John as a disgrace to the cloth; though he sometimes comforted himself with the reflection that most likely that very imaginative knight never existed. But he thoroughly enjoyed these Icelandic experiences, for, to use one of his own phrases, the power of the hills was upon him. With Mr. Lock he visited the concession, and on his way passed through a village where there was a fair, and where he had a very narrow escape. A little more, we are told, and a hideous, snuffy, old Icelandic woman would have kissed him. In respect to the survey, the mass of workable material was enormous. There was no lack of sulphur, and the speculation promised to be a remunerative one. Eventually, however, it was found that the obstacles were insuperable, and the scheme had to be abandoned. However, the trip had completed the cure commenced by Camoens, and at the end of it everybody said "he looked at least fifteen years younger."

Burton had scarcely left Granton for Iceland before Mrs. Arundell died, and the letters which Mrs. Burton wrote at this time throw an interesting light on the relations between her and Burton's family. To Miss Stisted she says (June 14th), "My darling child. My dear mother died in my arms at midnight on Wednesday 5th. It was like a child going to sleep, *most happy*, but quite unexpected by *us*, who thought, though sinking, she would last till August or October. I need not tell you, who know the love that existed between her and me, that my loss is bitter and irreparable, and will last for life. May *you* never know it! I have written pages full of family detail to darling Nana, and I intended to enclose it to you to read *en route*, but I thought perhaps our religious views and observances might seem absurd to the others.

and I felt ashamed to do so. You know when so holy a woman as dear mother dies, we do not admit of any melancholy or sorrow except for ourselves. Your dear little letter was truly welcome with its kind and comforting messages. I am glad that our darling [Burton] was spared all the sorrow we have gone through, and yet sorry he did not see the beauty and happiness of her holy death. . . She called for Richard twice before her death. Do write again and often, dear child. Tell me something about the Iceland visits. . . . Your loving Zooey."

What with the unsatisfactory condition of their affairs, and the death of her mother, Mrs. Burton was sadly troubled; but the long lane was now to have a turning. One day, while she was kneeling with wet cheeks before her mother's coffin, and praying that the sombrous overhanging cloud might pass away, a letter arrived from Lord Granville offering her husband the Consulate of Trieste [1] with a salary of £700 a year. This was a great fall after Damascus, but in her own words, "better than nothing," and she at once communicated with her husband, who was still in Iceland.

She then made a round of country house visits, including one to Wardour Castle.[2] In an unpublished letter to Miss Stisted, she says: "My pet, I came here on Tuesday. . . I have never cried nor slept since mother died (a month to-morrow) I go up again on Monday for final pack-up—to my convent ten days— . . . . then back to town in hopes of Nana in August, about the 7th. Then we shall go to Spain, and to Trieste, our new appointment, if he [Burton] will take it, as all our friends and relations wish, if only as a stop-gap for the present. Arundell has done an awfully kind thing. There is a large Austrian

70. Wardour Castle, 5th July 1872.

[1] It had just been vacated by the death of Charles Lever, the novelist. Lever had been Consul at Trieste from 1867 to 1872. He died at Trieste, 1st June 1872.

[2] Near Salisbury.

honour in the family with some privileges, and he has desired me to assume all the family honours on arriving, and given me copies of the Patent, with all the old signatures and attested by himself. This is to present to the Herald's College at Vienna. He has desired my cards to be printed Mrs. Richard Burton, née Countess Isabel Arundell of Wardour of the most sacred Roman Empire. This would give us an almost royal position at Vienna or any part of Austria, and with Nana's own importance and fame we shall (barring salary) cut out the Ambassador. She wants a quiet year to learn German and finish old writings. . . . I should like the tour round the world enormously, but I don't see where the money is to come from. . . This is such a glorious old place. . . The woods and parks are splendid, and the old ruin of the castle defended by Lady Blanche is the most interesting thing possible. Half the other great places I go to are mushroom greatness, but this is the real old thing of Druid remains and the old baronial castle of knights in armour and fair Saxon-looking women, and with heavy portcullises to enter by, and dungeons and subterranean passages, etc. There is a statue of our Saviour over the door, and in Cromwell's siege a cannon ball made a hole in the wall just behind it and never took off its head. . . Your loving Zoo."

A few days later Mrs. Burton received a letter from her husband, who expressed his willingness to accept Trieste. He arrived at Edinburgh again on September 5th, and his presence was the signal for a grand dinner, at which all the notables of the neighbourhood, including many people of title, were present. But, unfortunately, Burton was in one of his disagreeable moods, and by the time dinner was half over, he found that he had contradicted with acerbity every person within earshot. While. however, he was thus playing the motiveless ogre, his brother-in-law, Sir Henry Stisted, at the other end of the table, was doing his utmost to render himself agreeable,

DR. E. J. BURTON

MRS. E. J. BURTON.

and by the extraordinary means of rolling out anecdote after anecdote that told against the Scotch character. The Mackenzies, the Murrays, the MacDonalds, the McQueens, looked black as thunder, and Stisted's amiability gave even more offence than Burton's ill-temper. Noticing that something was amiss opposite him, Burton stopped his own talk to listen. Then Stisted's innocence and the ludicrousness of the whole scene dawned upon him, and leaning back in his chair he roared with uncontrollable laughter. When he met his wife again one of her first questions was about this dinner, at which she had hoped her husband would dazzle and delight the whole company, and which she supposed might lead to his promotion, He then told her the whole story, not omitting his ill-humour. She listened with dismay, and then burst into tears. "Come," he commented, "I wasn't so bad as Stisted, anyhow."

71. St. George and Frederick Burton.

Upon his return to London, Burton renewed his acquaintance with his cousins Dr. and Mrs. Edward John Burton. He and Dr. Burton, whom he thought fit to call after a character in *The Arabian Nights*, "Abu Mohammed Lazybones,"[1] had long known each other, but Dr. Burton had also for some time resided in distant lands. The notes that brought about the meeting—and they could not well be briefer—now lie before me, They run :

"Athenæum Club,
"Sept. 20 '72.

"My dear Cousin,
"When and where can I see you? Yours truly,
"R. F. Burton."

"Junior United Service Club.

"My dear Richard,
"Any day at 4 p.m.
"Yours ever,
"E. J. Burton."

(1) Burton's A.N. iv. Lib. Ed., iii., 282. Payne's A.N. iii., 10.

A few days later, Burton dined with Edward John, and made the acquaintance of his young cousins, St. George and Frederick. Of St. George, a dark-haired lad, who was particularly clever and had a humorous vein, Burton from the first thought highly. One day, happening to turn over some of the leaves of the boy's exercise book, he stumbled upon the following lines :

> "The map of Africa was dark as night,
> God said, 'Let Burton live,' and there was light."

He laughed heartily, and thanked his little cousin for the compliment, while the couplet became a stock quotation in the family. Later, when St. George went to a French school, he was very proud to find that the boys were conversant not only with the exploits of his famous uncle, but also with the history of the Dr. Francis Burton who had made Napoleon's death mask. Frederick Burton was a plump, shy, fair-haired little fellow, and Burton, who loved to tease, did not spare his rotundity. In one of Frederick's copy-books could be read, in large hand,

"Life is short."

"I," commented Burton, "find life very long."

Subsequently he advised his cousin to go to the River Plate. "Well," he would ask, when he entered the house, "has Frederick started for the River Plate yet? I see a good opening there."

As Dr. Burton was born in the house of his father's brother, the Bishop of Killala, Burton used to affect jealousy. "Hang it all, Edward," he would say, "You were born in a bishop's palace."

Apparently it was about this time that the terrible silence of Burton's brother was for a moment broken. Every human device had been tried to lead him to conversation, and hitherto in vain. It seems that some years previous, and before Edward's illness, Dr. E. J. Burton had lent his cousin a small sum of money, which was duly repaid. One day Dr. Burton chose to assume the contrary, and coming upon Edward suddenly he cried:

MRS. SEGRAVE.

"MY DEAR LOUISA."

"Edward, you might just as well have paid me that money I lent you at Margate. I call it shabby, now."

Edward raised his head and fixing his eyes on Dr. Burton said, with great effort, and solemnly, "Cousin, I did pay you, you must remember that I gave you a cheque."

Thrilled with joy, Dr. Burton attempted to extend the conversation, but all in vain, and to his dying day Edward Burton never uttered another word.

72. At the Athenæum.

Of all the spots in London, none was so dear to Burton as his Club, *The Athenæum*. When in England, he practically lived there, and its massive portico, its classic frieze, and the helmeted statue of Minerva were always imaged on his heart. He wrote a number of his books there, and he loved to write his letters on its notepaper stamped with the little oval enclosing Minerva's head. He used to make his way to the *Athenæum* early in the day[1] and go straight to the library. Having seated himself at the round table he would work with coralline industry, and without a single break until six or seven in the evening. It was a standing joke against him in Dr. Burton's family that when at the club he was never at home to anybody except a certain Mrs. Giacometti Prodgers. This lady was of Austrian birth, and, according to rumour, there was a flavour of romance about her marriage. It was said that while the laws of certain countries regarded her as married, those of other countries insisted that she was still single. However, married or not, she concentrated all her spleen on cab-drivers, and was continually hauling some luckless driver or other before the London magistrates. Having a profound respect for Burton's judgment, she often went to him about these cab disputes, and, oddly enough, though nobody else could get at him, he was always at the service of Mrs. Prodgers, and

(1) Told me by Mr. Henry Richard Tedder, librarian at the Athenæum from 1874.

good-naturedly gave her the benefit of his wisdom.[1] To the London magistrates the good lady was a perpetual terror, and Frederick Burton, a diligent newspaper reader, took a pleasure in following her experiences. "St. George," he would call across the breakfast table, "Mrs. Giacometti Prodgers again: She's had another cab-man up."

One evening, says a London contributor to the *New York Tribune*[2] referring to this period, "there was a smoking party given by a well-known Londoner. I went in late, and on my way upstairs, stumbled against a man sitting on the stairs, with a book and pencil in his hands, absorbed in his reading, and the notes he was making. It was Burton. When I spoke to him he woke up as if from a dream with the dazed air of one not quite sure where he is. I asked him what he was reading. It proved to be Camoens, and he told me he was translating the Portuguese poet. It seemed an odd place for such work, and I said as much. "Oh," answered Burton, "I can read anywhere or write anywhere. And I always carry Camoens about with me. You see, he is a little book, and I have done most of my translating in these odd moments, or, as you say, in this odd fashion. And he added, with a kind of cynical grin on his face, 'You will find plenty of dull people in the rooms above.' He had been bored and this was his refuge."

73. Jane Digby Again.

Report now arrived that Jane Digby was dead; and paragraphs derogatory to her character appeared in the press. Mrs. Burton not only answered them, but endeavoured to throw a halo over her friend's memory. She said also that as she, Mrs. Burton, had Jane Digby's biography, nobody else had any right to make remarks. Comically enough, news then came that Jane was still alive. She had been detained in the desert by the fighting

(1) Burton, who was himself always having disputes with cab-drivers and everybody else, probably sympathised with Mrs. Prodgers' crusade.

(2) Of 2nd November 1891.

GEORGE BURTON

FREDERICK BURTON.

of the tribes. Says Mrs. Burton, "her relatives attacked her for having given me the biography, and she, under pressure, denied it in print, and then wrote and asked me to give it back to her; but I replied that she should have had it with the greatest pleasure, only she having 'given me the lie' in print, I was obliged for my own sake to keep it, and she eventually died." This very considerate act of Jane's saved all further trouble.

On his expedition with Speke to Tanganyika, Burton had already written four volumes,[1] and it was now to be the subject of another work, *Zanzibar*, which is chiefly a description of the town and island from which the expedition started. The origin of the book was as follows. With him on his way home from Africa he had brought among other MSS. a bundle of notes relating both to his "preliminary canter" and to Zanzibar, and the adventures of these notes were almost as remarkable as those of the Little Hunchback. On the West Coast of Africa the bundle was "annexed" by a skipper. The skipper having died, the manuscripts fell into the hands of his widow, who sold them to a bookseller, who exposed them for sale. An English artillery officer bought them, and, in his turn, lost them. Finally they were picked up in the hall of a Cabinet Minister, who forwarded them to Burton. The work contains an enormous mass of geographical, anthropological and other information, and describes the town so truthfully that nobody, except under compulsion, would ever dream of going there. The climate, it seems, is bad for men, worse for women. "Why," he asks, "should Englishmen poison or stab their wives when a few months at Zanzibar would do the business more quietly and effectually?" The expense of getting them over may be one objection. But whoever goes to Zanzibar, teetotallers, we are told, should keep away. There it is drink or die. Burton

74. His Book on Zanzibar.

(1) *Lake Regions of Equatorial Africa* (2 vols. 1860). Vol. 33 of the *Royal Geographical Society*, 1860, and *The Nile Basin*, 1864.

introduces many obsolete words, makes attacks on various persons, and says fearlessly just what he thinks; but the work has both the Burtonian faults. It is far too long, and it teems with uninteresting statistics.

There also left the press this year (1872) a work in two volumes entitled *Unexplored Syria*, by Burton and Tyrwhitt Drake.[1] It describes the archæological discoveries made by the authors during their sojourn in Syria, and includes an article on Syrian Proverbs (*Proverba Communia Syriaca*) which had appeared the year before in the Journal of the Royal Asiatic Society. Some of the sayings have English analogues, thus:

"He who wants *nah*
Mustn't say *ah*;"

nah being wealth or honour; *ah*, the expression of fear or doubt.[2]

"He who to his neighbour's kind
Good in his own home will find."

"If Yais is your cook, pray eat,
But do not look for well-cooked meat."

At one of the meetings of the Royal Geographical Society, at which Burton had been billed to speak, there were present among the audience his wife, Mr. Arundell, and several other members of the family. Considerable hostility was shown towards Burton; and Colonel Rigby[3] and others flatly contradicted some of his statements respecting Zanzibar. Then Burton flew into a temper such as only he could fly into. His eyes flashed, his lips protruded with rage, and he brandished the long map pointer so wildly that the front bench became alarmed for their safety. Old Mr. Arundell, indignant at hearing his son-in-law abused, then tried to struggle on to the platform, while his sons and daughters, horrified at the prospect, hung like bull-dogs to his coat tails. Says

(1) A portion was written by Mrs. Burton.

(2) These are words used by children. *Unexplored Syria*, i., 288. *Nah* really means sweetstuff.

(3) Afterwards Major-General. He died in April 1887. See Chapter ix., § 38.

Burton, "the old man, who had never been used to public speaking, was going to address a long oration to the public about his son-in-law, Richard Burton. As he was slow and very prolix, he would never have sat down again, and God only knows what he *would* have said." The combined efforts of the Arundell family, however, prevented so terrible a denouement, Burton easily proved his enemies' statements to be erroneous, and the order was eventually restored.

chiefly women. "Trieste," said Burton, "is a town of threes—three quarters, three races (Italian, Slav and Austrian), and three winds (Sirocco, Bora, and Contraste)." One brilliant man of letters had been connected with the town, namely Marie-Henri Beyle, better known by his pen name, Stendhal,[1] who, while he was French Consul here, pumice polished and prepared for the press his masterpiece, *La Chartreuse de Parme*, which he had written at Padua in 1830. To the minor luminary, Charles Lever, we have already alluded. Such was the town in which the British Hercules was set to card wool. The Burtons occupied ten rooms at the top of a block of buildings situated near the railway station. The corridor was adorned with a picture of our Saviour, and statuettes of St. Joseph and the Madonna with votive lights burning before them. This, in Burton's facetious phrase, was "Mrs. Burton's joss house;" and occasionally, when they had differences, he threatened "to throw her joss house out of the window." Burton in a rage, indeed, was the signal for the dispersal of everybody. Furniture fell, knick-knacks flew from the table, and like Jupiter he tumbled gods on gods. If, however, he and his wife did not always symphonize, still, on the whole, they continued to work together amicably, for Mrs. Burton took considerable pains to accommodate herself to the peculiarities of her husband's temperament, and both were blessed with that invaluable oil for troubled waters—the gift of humour. "Laughter," Burton used to say, and he had "a curious feline laugh," "animates the brain and stimulates the lungs." To his wife's assumption of the possession of knowledge, of being a linguist, of being the intellectual equal of every living person, saving himself, he had no objection; and the pertinacity with which she sustained this rôle imposed sometimes even on him. He got to think that she was really a genius in a

(1) Stendhal, born 1783. Consul at Trieste and Civita Vecchia from 1830 to 1839. Died in Paris, 23rd March 1842. Burton refers to him in a footnote to his Terminal Essay in the *Nights* on "Al Islam."

way, and saw merit even in the verbiage and rhodomontade of her books. But whatever Isabel Burton's faults, they are all drowned and forgotten in her devotion to her husband. It was more than love—it was unreasoning worship. "You and Mrs. Burton seem to jog along pretty well together," said a friend. "Yes," followed Burton, "I am a spoilt twin, and she is the missing fragment."

Burton, of course, never really took to Trieste, his Tomi, as he called to it. He was too apt to contrast it with Damascus: the wind-swept Istrian hills with the zephyr-ruffled Lebanon, the dull red plains of the Austrian sea-board with the saffron of the desert, the pre-historic castellieri or hill-forts, in which, nevertheless, he took some pleasure, with the columned glories of Baalbak and Palmyra. "Did you like Damascus?" somebody once carelessly asked Mrs. Burton.

"Like it!" she exclaimed, quivering with emotion, "My eyes fill, and my heart throbs even at the thought of it."

Indeed, they always looked back with wistful, melancholy regret upon the two intercalary years of happiness by the crystalline Chrysorrhoa, and Mrs. Burton could never forget that last sad ride through the beloved Plain of Zebedani. Among those who visited the Burtons at Trieste, was Alfred Bates Richards. After describing Mrs. Burton's sanctuary, he says: "Thus far, the belongings are all of the cross, but no sooner are we landed in the little drawing-rooms than signs of the crescent appear. These rooms, opening one into another, are bright with Oriental hangings, with trays and dishes of gold and burnished silver, fantastic goblets, chibouques with great amber mouth-pieces, and Eastern treasures made of odorous woods." Burton liked to know that everything about him was hand made. "It is so much better," he used to say, than the "poor, dull work of machinery." In one of the book-cases was Mrs. Burton's set of her husband's works, some fifty volumes.[1]

(1) These are all preserved now at the Central Library, Camberwell.

Mr. Richards thus describes Burton himself, "Standing about five feet eleven, his broad, deep chest and square shoulders reduce his apparent height very considerably, and the illusion is intensified by hands and feet of Oriental smallness. The Eastern and distinctly Arab look of the man is made more pronounced by prominent cheek-bones (across one of which is the scar of a javelin cut), by closely-cropped black hair, just tinged with grey, and a pair of piercing, black, gipsy-looking eyes." Out of doors, in summer, Burton wore a spotlessly white suit, a tie-pin shaped like a sword, a pair of fashionable, sharply-pointed shoes, and the shabbiest old white beaver hat that he could lay his hands upon. On his finger glittered a gold ring, engraved with the word "Tanganyika."[1] In appearance, indeed, he was a compound of the dandy, the swash-buckler and the literary man. He led Mr. Richards through the house. Every odd corner displayed weapons—guns, pistols, boar-spears, swords of every shape and make. On one cupboard was written "The Pharmacy." It contained the innocuous medicines for Mrs. Burton's poor—for she still continued to manufacture those pills and drenches that had given her a reputation in the Holy Land. "Why," asked Richards, "do you live in a flat and so high up?" "To begin with," was the reply, "we are in good condition, and run up and down the stairs like squirrels. If I had a great establishment, I should feel tied and weighed down. With a flat and two or three servants one has only to lock the door and go out." The most noticeable objects in the rooms were eleven rough deal tables, each covered with writing materials.[2] At one sat Mrs. Burton in morning négligé, a grey choga—the long, loose Indian dressing-gown of soft camel's hair—topped by a smoking cap of the same material. She observed, "I see you are looking at our tables. Dick likes a separate table for each book, and when he is tired of one he goes

(1) Now in the possession of Mrs. St. George Burton.

(2) In later times Dr. Baker never saw more than three tables.

to another." He never, it seems, wrote more than eleven books at a time, unless stout pamphlets come under that category. Their life was a peaceful one, except on Fridays, when Mrs. Burton received seventy bosom and particular friends, and talked to them at the top of her voice in faulty German, Italian, which she spoke fluently, or slangy English.[1] In the insipid conversation of this "magpie sanhedrin," "these hen parties," as he called them, Burton did not join, but went on with his work as if no one was present. Indeed, far from complaining, he remarked philosophically that if the rooms had been lower down probably 140 visitors instead of 70 would have looked in. The Burtons usually rose at 4 or 5, and after tea, bread and fruit, gave their morning to study. At noon they drank a cup of soup, fenced, and went for a swim in the sea. Burton then took up a heavy iron stick with a silver knob[2] and walked to the Consulate, which was situated in the heart of the town, while Mrs. Burton, with her pockets bulging with medicines, and a flask of water ready for baptism emergencies hanging to her girdle, busied herself with charitable work, including the promotion of the Society for the Prevention of Cruelty to Animals. They generally dined at the table d'hôte of the Hotel de la Ville, and dined well, for, as Burton says used to "Only fools and young ladies care nothing for the *carte*."[3] Having finished their coffee, cigarettes, and *kirsch*, outside the hotel, they went home to bed, where, conscious of a good day's work done, they took their rest merrily. Sometimes they interrupted the routine with excursions into the surrounding country, of which they both knew every stock and stone, pre-historic or modern. Of business ability, Burton had never possessed one iota, and his private

(1) Mrs. Burton, was, of course, no worse than many other society women of her day. Her books bristle with slang.

(2) It is now in the possession of Mrs. E. J. Burton, 31, Wilbury Road, Brighton.

(3) Later Burton was himself a sad sinner in this respect. His studies made him forget his meals.

affairs were constantly mis-managed. As at Fernando Po, Santos and Damascus, he promptly looked out for a sanitarium, his choice finally resting upon a loftily-situated village called Opçina.

Reviewing Burton's carreer, Mr. Alfred Bates Richards says: "He has done more than any other six men, and is one of the best, noblest and truest that breathes. While not on active service or on sick leave he has been serving his country, humanity, science, and civilisation in other ways, by opening up lands hitherto unknown, and trying to do good wherever he went. He was the pioneer for all other living African travellers."

If Trieste was not an ideal post for him, still it had the patent advantage of being practically a sinecure. He and his wife seem to have been able to get away almost at any time. They sometimes travelled together, but often went in different directions, and as Burton was as restless as a hyena, he never stayed in any one place many hours. Occasionally they met unexpectedly. Upon one of these meetings in a Swiss hotel, Burton burst out affectionately with: "And what the devil brought *you* here?" To which she replied, promptly but sweetly, "Ditto, brother." For study, Burton had almost unlimited time, and nothing came amiss to him. He lost himself in old sacramentaries, Oriental manuscripts, works on the prehistoric remains of Istria, Camoens, Catullus, *The Arabian Nights*, Boccaccio. His knowledge was encyclopædic.

Early in 1873 the Burtons visited Vienna chiefly in order to see the great Exhibition. The beauty of the buildings excited their constant admiration, but the dearness of everything at the hotels made Burton use forcible language. On one occasion he demanded—he never asked for anything—a beefsteak, and a waiter hurried up with an absurdly small piece of meat on a plate. Picking it up with the fork he examined it critically, and then

76. At the Vienna Exhibition, 1873.

said, quite amiably for him, "Yaas, yaas,[1] that's it, bring me some." Next he required coffee. The coffee arrived in what might have been either a cup or a thimble. "What's this?" demanded Burton. The waiter said it was coffee for one. "Then," roared Burton, with several expletives, "bring me coffee for twenty." Their bill at this hotel came to £163 for the three weeks.

77. A Visit from Drake, June 1873.

On their return from Vienna, they had the pleasure of meeting again Lady Marion Alford, Aubertin, and that "true-hearted Englishman, staunch to the backbone," Charles Tyrwhitt Drake, who "brought with him a breath from the desert and stayed several weeks." The three friends went to a fête held in the stalactite caverns of Adelsberg, from which Burton, who called them the eighth wonder of the world, always assumed that Dante got his ideas of the Inferno. Lighted by a million candles, and crowded with peasants in their picturesque costumes, which made wondrous arabesques of moving shadows, the caves presented a weird and unearthly appearance, which the music and dancing subsequently intensified. Shortly afterwards Drake left for Palestine. In May (1874), Burton was struck down by a sudden pain, which proved to arise from a tumour. An operation was necessary, and all was going on well when a letter brought the sad news of Drake's death. He had succumbed, at Jerusalem, to typhoid fever, at the early age of twenty-eight.[2] Burton took the news so heavily, that, as Mrs. Burton says,[3] it "caused the wound to open afresh; he loved Drake like a brother, and few know what a tender heart Richard has." To use Dr Baker's[4] phrase, he had "the heart of a beautiful woman."

(1) His usual pronunciation of the word.

(2) 12th August 1874.

(3) Letter to Lord Houghton.

(4) Dr. Grenfell Baker, afterwards Burton's medical attendant.

In the meantime Mrs. Burton was reaping the fruits of her injudicious treatment of Khamoor. Thoroughly spoilt, the girl now gave herself ridiculous airs, put herself on a level with her mistress, and would do nothing she was told. As there was no other remedy, Mrs. Burton resolved philanthropically to send her back to Syria, "in order that she might get married and settled in life." So Khamoor was put on board a ship going to Beyrout, with nine boxes of clothes and a purse of gold. "It was to me," says Mrs. Burton, "a great wrench." Khamoor's father met her, the nine boxes, and the purse of gold at Beyrout, and by and by came the news that she was married and settled down in the Buka'a. Such was the end of Chico the Second.

78. Khamoor returns to Syria, 4th December 1874.

CAPTAIN BURTON.

FROM THE PORTRAIT BY (SIR) FREDERICK LEIGHTON NOW IN THE NATIONAL PORTRAIT GALLERY.

Photograph lent by Mr. Augustin Rischgitz.

CHAPTER XVIII

12TH MAY 1875—18TH JUNE 1876

THE TRIP TO INDIA

BIBLIOGRAPHY:

51. The Port of Trieste.
52. The Gypsy. Written in 1875.
53. Etruscan Bologna. 1876.
54. New System of Sword Exercise for Infantry. 1876.

On 8th December 1874, Burton sent his wife to England to arrange for the publication of various of his works, and in May 1875, having obtained leave, he followed her, arriving in London on the 12th. He took with him "a ton or so of books" in an enormous trunk painted one half black the other white—"the magpie chest" which henceforth always accompanied him on his travels. At the various stations in England there were lively scenes, the company demanding for luggage excess, and Burton vigorously protesting but finally paying. He then took the value out by reeling off a spirited address to the railway clerk, punctuated with expletives in twenty odd African or Asiatic languages, on the meanness of the clerk's employers.

79. Visit to England, 12th May 1875.

Always suffering from impecuniosity, the Burtons were perpetually revolving schemes for increasing their income. One was to put on the market a patent pick-me-up, good also for the liver, to be called, "Captain Burton's Tonic Bitters," the recipe of which had been "acquired from a Franciscan monk." "Its object," observed Burton

80. Tonic Bitters.

facetiously, to a friend, "is to make John Bull eat more beef and drink more beer." Mrs. Burton imagined naïvely that if it were put into a pretty bottle the demand would exceed the supply. They had hopes, too, for the Camoens, which had taken many years of close application and was now approaching completion. Still, it was argued that a Translation of Camoens, however well done, could not hope for the success of a well-advertised liver tonic, seeing that while most people have a liver, it is only here and there one who has a taste for Camoens. The tonic was placed on the market, but the scheme, like so many others, proved a fiasco. Nobody seemed to want to be picked up, and the indifference of a Christian nation to the state of its liver, was to Burton extremely painful. So he abandoned philanthropy, and took to lecturing before the Anthropological and other societies, dining out, and calling on old friends. One Sunday he visited the Zoo; but when he asked for a glass of beer at the refreshment bar, the girl declined to serve him because he was "not a *bonâ-fide* traveller!"

In 1875, Burton's portrait, painted by Sir Frederick Leighton, was exhibited in the Academy; and on July 6th of the same year, Burton started off on a second trip to Iceland, which occupied him six weeks, but he and his wife did not meet again till October 6th. On December 4th (1875) they left London for the Continent. The morning was black as midnight. Over the thick snow hung a dense, murky fog, while "a dull red gleam just rendered the darkness visible."

"It looks," said Burton, "as if London were in mourning for some great national crime."

To which Mrs. Burton replied, "Let us try to think, darling, that our country wears mourning for our departure into exile."

On reaching Boulogne they sought out some of their old acquaintances, including M. Constantin, Burton's fencing master. After a brief stay in Paris, they proceeded to Trieste, ate their Christmas dinner, and then set out for India, partly for pleasure and partly for the purpose of collecting information about the abandoned diamond mines of Golconda.

81. A Trip to India, December 1875, 18th June 1876.

The Suez Canal, which had been finished some five years previous, gave them much pleasure, and it was like living life over again to see the camels, the Bedawin in cloak and kuffiyyah, the women in blue garments, and to smell the pure air of the desert. On reaching Yambu, Burton enquired whether Sa'ad the robber chief, who had attacked the caravan in the journey to Mecca days, still lived; and was told that the dog had long since made his last foray, and was now safe in Jehannum.(1) They landed at Jiddah, where Burton was well received, although everyone knew the story of his journey to Mecca, and on rejoining their ship they found on board eight hundred pilgrims of a score of nationalities. Then a storm came on. The pilgrims howled with fright, and during the voyage twenty-three died of privation, vermin, hunger and thirst. Says Mrs. Burton:(2) "They won't ask, but if they see a kind face they speak with their eyes as an animal does." At Aden Burton enquired after his old Harar companions. Shahrazad was still in Aden, the coquettish Dunyazad in Somaliland, the Kalandar had been murdered by the Isa tribe, and The End of Time had "died a natural death"—that is to say, somebody had stuck a spear into him.(3) Bombay was reached on February 2nd.

82. Arbuthnot Again. Rehatsek.

The first person Burton called on was his old friend, Forster FitzGerald Arbuthnot, who now occupied there the important position of "Collector." Arbuthnot, like other people, had got older, but his character had not changed a tittle. Business-like and shrewd, yet he continued to be kindly, and would go out of his way to do a

(1) Hell.

(2) *A.E.I.* (Arabia, Egypt, India).

(3) Burton's A.N., v., 304. Lib. Ed., vol. 4, p. 251.

philanthropic action, and without fuss or parade. A friend describes him as "a man of the world, but quite untainted by it." He used to spend the winter in Bombay, and the summer in his charming bungalow at Bandora. In a previous chapter we referred to him as a Jehu. He now had a private coach and team—rather a wonder in that part of the world, and drove it himself. Of his skill with the ribbons he was always proud, and no man could have known more about horses. Some of the fruits of his experience may be seen in an article[1] which he contributed to *Baily's Magazine* (April 1883) in which he ranks driving with such accomplishments as drawing, painting and music. His interest in the languages and literatures of the East was as keen as ever, but though he had already collected material for several books he does not seem to have published anything prior to 1881. He took his friends out overywhere in his four-in-hand, and they saw to advantage some of the sights of Burton's younger days. With the bungalow Mrs. Burton was in raptures. On the eve of the Tabút feast, she tells us, the Duke of Sutherland (formerly Lord Stafford) joined the party; and a number of boys dressed like tigers came and performed some native dancing with gestures of fighting and clawing one another, "which," she adds oddly, "was exceedingly graceful."

The principal event of this visit, however, was Burton's introduction to that extraordinary and Diogenes-like scholar, Edward Rehatsek. Lady Burton does not even mention Rehatsek's name, and cyclopædias are silent concerning him; yet he was one of the most remarkable men of his time, and henceforward Burton was in constant communication with him. Born on 3rd July 1819, at Illack, in Austria, Edward Rehatsek was educated at Buda Pesth, and in 1847 proceeded to Bombay, where he settled down as Professor of Latin and mathematics at Wilson College. He retired from his professorship in 1871, and settled in a reed-built native house, not so very much bigger

(1) About driving four horses.

than his prototype's tub, at Khetwadi. Though he had amassed money he kept no servants, but went every morning to the bazaar, and purchased his provisions, which he cooked with his own hand. He lived frugally, and his dress was mean and threadbare, nevertheless, this strange, austere, unpretentious man was one of the greatest linguists of his time. Not only could he speak most of the languages of the East, including Arabic and Persian, but he wrote good idiomatic English. To his translations, and his connection with the Kama Shastra Society, we shall refer later. He was visited in his humble home only by his principal friend, Mr. Arbuthnot, and a few others, including Hari Madhay Parangpe, editor of *Native Opinion*, to which he was a contributor. The conversation of Rehatsek, Burton, and Arbuthnot ran chiefly on Arbuthnot's scheme for the revival of the Royal Asiatic Translation Fund, and the translation of the more important Eastern works into English; but some years were to elapse before it took shape.

On February 4th, Burton wrote to his cousin, St. George Burton—addressing his letter, as he was continually on the move, from Trieste. He says:

"My Dear Cousin,

"You need not call me 'Captain Burton.' I am very sorry that you missed Woolwich—and can only say, don't miss the Line. I don't think much of Holy Orders, however, *chacun à son goût*. Many thanks for the details about the will. Assist your mother in drawing up a list of the persons who are heirs, should the girl die without a will.[1] Let 'the party' wash his hands as often as he pleases—cleanliness is next to godliness. As the heir to a baronetcy[2] you would be worth ten times more than heir to an Esquireship—in snobby England. Write to me whenever you think that I can be of any service and let me be "Yr. aff. cousin,

"R. F. Burton."

(1) I do not know to what this alludes.

(2) See Chapter i.

From Bombay, the Burtons journeyed to Karachi, which had grown from 3,000 to 45,000,[1] and could now boast fine streets and noble houses. Here Burton regaled his eyes with the sights familiar to his youth; the walks he had taken with his bull-terrier, the tank or pond where he used to charioteer the "ghastly" crocodile,[2] the spot where he had met the beautiful Persian, and the shops which had once been his own; while he recalled the old familiar figures of hook-nosed Sir Charles Napier, yellow-bearded Captain Scott, and gorgeously-accoutred General J-J-J-J-J-J-Jacob. His most amusing experience was with a Beloch chief, one Ibrahim Khan, on whom he called and whom he subsequently entertained at dinner spread in a tent.[3] The guest, Sind fashion, prepared for the meal by getting drunk. He thoroughly enjoyed it, however, and, except that he made impressions with his thumb in the salt, upset his food on the tablecloth, and scratched his head with the corkscrew, behaved with noticeable propriety. Having transferred from the table to his pocket a wine-glass and some other little articles that took his fancy, he told his stock stories, including the account of his valour at the battle of Meeanee, where at imminent risk of his life, he ran away. Tea he had never before tasted, and on sampling a cup, he made a wry face. This, however, was because it was too strong, for having diluted it with an equal quantity of brandy, he drank it with relish.

83. In Sind.

After a visit to the battlefield of Meeanee[4] the Burtons returned to Bombay in time for the feast of Muharram, and saw the Moslem miracle play representing the martyrdom and death of Hassan and Hossein, the sons of Ali. Then Mirza Ali Akbar, Burton's old munshi, called on them. As his visiting card had been printed

(1) Its population is now 80,000.

(2) *Sind Revisited*, i., 82.

(3) See *Sind Revisited*, vol. ii., pp. 109 to 149.

(4) Where Napier with 2,800 men defeated 22,000.

Mirza Ally Akbar, Burton enquired insultingly whether his old friend claimed kin with Ally Sloper. In explanation the Mirza said that the English were accustomed to spell his name so, and as he did not in the least mind what he was called, he had fallen in with the alteration.

84. Golconda.

On February 21st the Burtons left Bombay and journeyed by way of Poona to Hyderabad, where they were hospitably entreated by Major Nevill, the Commander-in-Chief of the Nizam's troops, and Sir Salar Jung, the Prime Minister. They rode through the town on elephants, saw the Nizam's palace, which was "a mile long and covered with delicate tracery," an ostrich race, an assault-at-arms, and fights between cocks and other creatures. At "Hyderabad," says Mrs. Burton, "they fight every kind of animal." "A nautch," which Sir Salah gave in their honour, Mrs. Burton found tame, for the girls did nothing but eat sweetmeats and occasionally run forward and twirl round for a moment with a half-bold, semi-conscious look.[1]

Then followed the visit to Golconda and its tombs of wax-like Jaypur marble, with their arabesqued cupolas and lacery in stone. Here Burton accumulated a good deal of miscellaneous information about diamond mining, and came to the conclusion that the industry in India generally, and especially in Golconda, had been prematurely abandoned; and endeavoured by means of letters to the press and in other ways to enlist the sympathies of the British capitalists. But everything that he wrote on the subject, as on kindred subjects, has a distinctly quixotic ring, and we fear he would not have been a very substantial pillar for the British capitalist to lean against. He was always, in such matters, the theorist rather than the practical man—in other words, the true son of his own father.

The Burtons then returned to Bombay, which they reached in time to take part in the celebrations in honour

(1) *Romance of Isabel Lady Burton*, ii., 584.

of the Prince of Wales, who had just finished his Indian tour. Honouring the Guebres—the grand old Guebres, as he used to call them—and their modern representatives, the Parsees, Burton paid a visit to the Parsee "burying place"—the high tower where the dead are left to be picked by vultures, and then he and his wife left for Goa, where they enjoyed the hospitality and company of Dr. Gerson Da Cunha,[1] the Camoens student and enthusiast.

Mrs. Burton was as disgusted with Goa as she had been charmed with Dr. Da Cunha. She says, "Of all the God-forgotten, deserted holes, one thousand years behind the rest of creation, I have never seen anything to equal it. They left India at the end of April, and were back again at Trieste on June 18th.

[1] Dr. Da Cunha, who was educated at Panjim, spent several years in England, and qualified at the Colleges of Physicians and Surgeons. He built up a large practice in Goa.

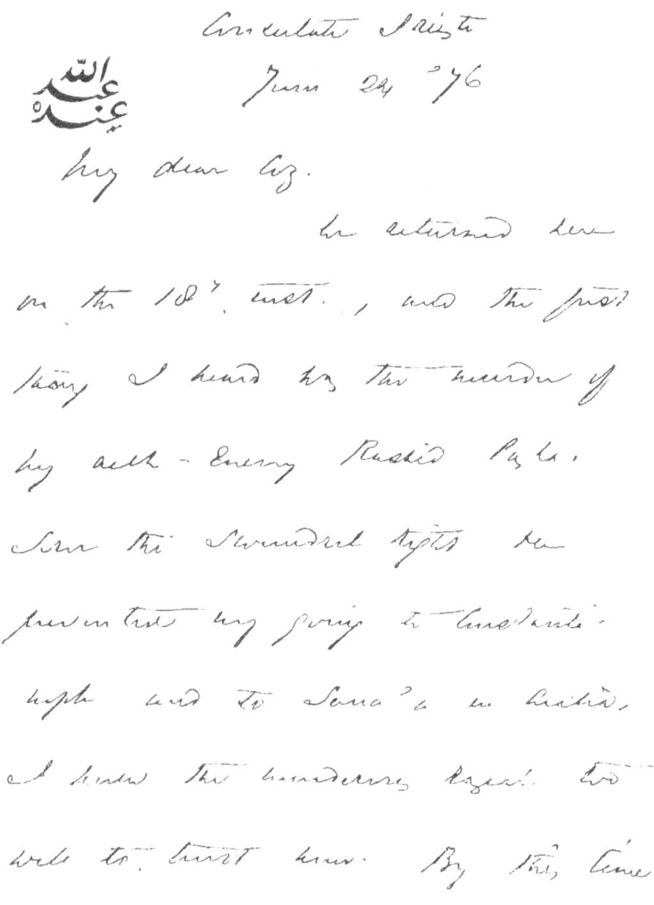

Consulate Trieste
June 24 '76

عبدالله عنده

My dear Coz.

We returned here on the 18th inst., and the first thing I heard was the murder of my old enemy Rashid Pasha. Since the scoundrel [illegible] has prevented my going to Constantinople and to Sana'a in Arabia, I knew the murdering rascal too well to trust him. By this time

PORTION OF BURTON'S LETTER TO DR. E. BURTON.

24TH JUNE, 1876.

CHAPTER XIX

18TH JUNE 1876—31ST MARCH 1877

COLONEL GORDON

Shortly after his return from India, Burton commenced a translation of the *Orlando Furioso* [1] of Ariosto, a poet, to whom, as we have seen, he had been drawn ever since those far-off days when with his father and the rest of the family he had meandered about Italy in the great yellow chariot. Reggio, the poet's birthplace, and Ferrara, where the *Orlando Furioso* was written and Ariosto died, were sacred spots to him; while the terrific madness of the hero, the loves of Ruggiero and Bradamante and the enchanted gardens with their Arabian Nights atmosphere, lapped him in bliss much as they had done in the old days. Only a small portion of this translation was ever finished, but he had it in mind all the rest of his life, and talked about it during his last visit to England.

85. Ariosto.

In June came the news of the murder of Rashid Pasha; and a thousand memories, sweet and bitter, thrilled the Burtons. Mrs. Burton recalled that "cool and aromatic housetop," the jewel-blue Chrysorrhoa, the saffron desert, and then it was "Oh, Rashid Pasha! Oh, Rashid Pasha!" Still she found it in her woman's heart to forgive the detested old enemy, now that he was

86. Death of Rashid Pasha, 24th June 1876.

(1) There are many English translations, from Harrington's, 1607, to Hoole's, 1783, and Rose's, 1823. The last is the best.

gone, but Burton could not restrain a howl of triumph such as might have become some particularly vindictive Bible hero.

Writing on 24th June to his cousin, Dr. Edward John Burton, he says, "We returned here on the 18th inst., and the first thing I heard was the murder of my arch-enemy, Rashid Pasha. Serve the scoundrel right. He prevented my going to Constantinople and to Sana'a, in Arabia. I knew the murderous rascal too well to trust him. Maria wrote to me about poor Stisted's death.[1] A great loss for Maria and the chicks. I suppose you never see Bagshaw.[2] What news are there of him? Is Sarah (What's her name? Harrison?)[3] still to the fore. It is, I fear, useless to write anything about poor Edward[4] except to thank you most heartily for your disinterested kindness to him. I will not bother you about our journey, which was very pleasant and successful. You will see it all, including my proposals for renewed diamond digging, written in a book or books."

"United best love to my cousin and the cousinkins."

Burton made frequent enquiries after Edward, "Many thanks," he writes on a post card, "for the news of my dear brother," and all his letters contain tender and warm-hearted references to him.

In July 1875, Burton heard from Colonel (afterwards General) Gordon, who wanted some information about the country south of the Victoria Nyanza; and the friendship which then commenced between these brilliant men was terminated only by death. In every letter Gordon quoted Burton's motto, "Honour, not honours," and in one he congratulated his friend on its happy choice. For several years Gordon had been occupied under the

87. Colonel Gordon 1877.

(1) Sir Henry Stisted died of consumption in 1876.

(2) Robert Bagshaw, he married Burton's aunt, Georgiana Baker.

(3) His cousin Sarah, who married Col. T. Pryce Harrison. See Chapter iv. and Chapter xix.

(4) Burton's brother.

auspices of the Khedive, in continuing the work of administering the Soudan, which had been begun by Sir Samuel Baker. He had established posts along the Nile, placed steamers on the Albert Nyanza, and he nursed the hope of being able to put an end to the horrid slave trade. In January 1877, he was appointed by the Khedive Governor of the entire Soudan. There were to be three governors under him, and he wrote to Burton offering him the governor-generalship of Darfur, with £1,600 a year. Said Gordon, "You will soon have the telegraph in your capital, El Fasher. . . . You will do a mint of good, and benefit those poor people. . . Now is the time for you to make your indelible mark in the world and in these countries."[1]

Had such an offer arrived eight years earlier, Burton might have accepted it, but he was fifty-seven, and his post at Trieste, though not an agreeable one, was a "lasting thing," which the governor-generalship of Darfur seemed unlikely to be. So the offer was declined. Gordon's next letter (27th June 1877) contains a passage that brings the man before us in very vivid colours. "I dare say," he observes, "you wonder how I can get on without an interpreter and not knowing Arabic. I do not believe in man's free will; and therefore believe all things are from God and pre-ordained. Such being the case, the judgments or decisions I give are fixed to be thus or thus, whether I have exactly hit off all the circumstances or not. This is my raft, and on it I manage to float along, thanks to God, more or less successfully."[2]

On another occasion Gordon wrote, "It is a delightful thing to be a fatalist"—meaning, commented Burton, "that the Divine direction and pre-ordination of all things saved him so much trouble of forethought and afterthought. In this tenet he was not only a Calvinist but also a Moslem."[3]

(1) *Romance of Isabel, Lady Burton*, ii., 656.

(2) *Romance of Isabel Lady Burton.*

(3) Burton's A.N., Suppl., ii., 61. Lib. Ed. ix., p. 286, note.

The patent Pick-me-up having failed, and the Burtons being still in need of money, other schemes were revolved, all more or less chimerical. Lastly, Burton wondered whether it would be possible to launch an expedition to Midian with a view to searching for gold. In ancient times gold and other metals had been found there in abundance, and remains of the old furnaces still dotted the country. Forty cities had lived by the mines, and would, Burton averred, still be living by them but for the devastating wars that had for centuries spread ruin and destruction. He, reasoned, indeed, much as Balzac had done about the mines of Sardinia as worked by the Romans, and from no better premises; but several of his schemes had a distinctly Balzacian aroma,(1) as his friend Arbuthnot, who was writing a life of Balzac, might have told him. Burton himself, however, had no misgivings. His friend, Haji Wali, had indicated, it seems, in the old days, the precise spot where the wealth lay, and apparently nothing remained to be done except to go and fetch it.

88. Jane Digby the Second.

Haji Wali had some excellent points. He was hospitable and good-natured, but he was also, as Burton very well knew, cunning and untrustworthy. The more, however, Burton revolved the scheme in his mind, the more feasible it seemed. That he could persuade the Khedive to support him he felt sure; that he would swell to bursting the Egyptian coffers and become a millionaire himself was also taken for granted, and he said half in earnest, half in jest, that the only title he ever coveted was Duke of Midian. There were very eager ears listening to all this castle building. At Trieste, Mrs. Burton had taken to her bosom another Jane Digby—a creature with soft eyes, "bought blushes and set smiles." One would have thought that former experiences would have made her cautious. But it was not so. Mrs. Burton,

(1) Thus, Balzac tried to discover perpetual motion, proposed to grow pineapples which were to yield enormous profits, and to make opium the staple of Corsica, and he studied mathematical calculations in order to break the banks at Baden-Baden.

though deplorably tactless, was innocence itself, and she accepted others at their own valuation. Jane Digby the Second, who went in and out of the Burton's house as if she belonged to it, was in reality one of the most abandoned women in Trieste. She was married, but had also, as it transpired, an acknowledged lover.

Like women of that class she was extravagant beyond belief, and consequently always in difficulties. Hearing the everlasting talk about Midian and its supposed gold, the depraved woman (1) made up her mind to try to detach Burton's affections from his wife and to draw them to herself. To accomplish this she relied not only on the attractions of her person, but also on glozing speeches and other feminine artifices. Having easy access to the house she purloined private letters, papers and other writings, and after all hope of recovery was over, she would put them back. She slipped love letters, purporting to be from other women, into Burton's pockets; and whenever Mrs. Burton brushed his coat or dried his clothes she was sure to come upon them Mrs. Burton also received pseudonymous letters.

But whatever Mrs. Burton's faults, she, as we have seen, passionately loved, trusted and even worshipped her husband; and whatever Burton's faults, he thoroughly appreciated her devotion. They were quite sufficient for each other, and the idea of anyone trying to come between them seemed ludicrous. Consequently Mrs. Burton carried her letters to her husband and he brought his to her. Amazing to say, neither of them suspected the culprit, though Burton thought it must be some woman's intrigue, and that need of money was the cause of it.

The real truth did not come out till after Burton's death, and then the unhappy woman, who was near her end, made Lady Burton a full confession, adding, "I

(1) We are telling the tale much as Mrs. Burton told it, but we warn the reader that it was one of Mrs. Burton's characteristics to be particularly hard on her own sex and also that she was given to embroidering.

took a wicked pleasure in your perfect trust in me.''

89. The Old Baronetcy. 18th January 1877.

Repeated enquiry now took place respecting the old baronetcy in the Burton family, and Mrs. Burton in particular made unceasing efforts, both in the columns of *Notes and Queries* and elsewhere, in order to obtain the missing links. Several of Burton's letters at this period relate to the subject. To Mrs. E. J. Burton, 18th January 1877, he writes: "My dear cousin, I write to you in despair: That 'party,' your husband, puts me off with a post-card, to this effect, 'Have seen W——ll, no chance for outsiders,' and does not tell me a word more. I wish you would write all you know about it. Another matter. Had the old man left me his money or any chance of it, I should have applied for permission to take up the old baronetcy. But now I shall not. Your husband is *the* baronet and he can if he likes assume the "Sir" at once. Why the devil doesn't he? Of course I advise him to go through the usual process, which will cost, in the case of a baronetcy, very few pounds. Neither he nor you may care for it, but think of the advantage it will be to your children. Don't blink the fact that the British public are such snobs that a baronet, even in the matrimonial market, is always worth £50,000, and it is one of the oldest baronetcies in the kingdom. Do take my advice and get it for your eldest son [St. George Burton]. As I said before, your husband might assume it even without leave, but he had better get 'the Duke' to sanction it. And don't fail to push the man, who won't even claim what is his right. *Que diable*! Am I the only article named Burton that has an ounce of energy in his whole composition.''

CHAPTER XX

31ST MARCH 1877 TO 27TH DECEMBER 1879

MIDIAN

BIBLIOGRAPHY:

55. Sind Revisited. 1877
56. The Gold Mines of Midian. 1878.
57. A.E.I. (Arabia, Egypt, India) by Isabel Burton. 1879.
58. Ogham Runes. 1879.
59. The Land of Midian Revisited. 2 vols., 1879.

90. "The New Joseph." 31st March 1877—21st April 1877. 19th October 1877—20th April 1878.

Burton now felt that the time was ripe to broach his views concerning the golden Chersonese to the Khedive (Ismail), and having easily obtained leave from the home authorities, he proceeded straight to Cairo. The Khedive, impressed with his representations and enthusiasm, promptly consented to supply funds, and "the New Joseph," as Burton was now called, began preparations for the expedition that was to make both Egypt and himself rich beyond computation. Then followed a conversation with Haji Wali, whom age—he was 77—"had only made a little fatter and a little greedier," and the specious old trickster promised to accompany the expedition. As usual Burton began with a preliminary canter, visiting Moilah, Aynunah Bay, Makna and Jebel Hassani, where he sketched, made plans, and collected metalliferous specimens. He returned to Egypt with native stories of ruined towns evidencing a formerly dense population, turquoise mines and rocks veined with gold. The Khedive in idea saw himself a second Crœsus. These were the quarries, he held, whence Solomon derived the gold for the walls of the house of his God, his drinking vessels and his lion throne, but Colonel Gordon, when afterwards told of the scheme, smiled incredulously. As the hot season necessitated a delay of six months, Burton returned to Trieste, where life seemed hum-drum enough after so many excitement, and spangled visions. He spent the time writing a book *The Gold Mines of Midian and the Ruined Midianite Cities*, and the sluggish months having at last crawled by, he again left Trieste for Cairo.

In a letter to Mrs. E. J. Burton, headed "At Sea, 8th May 1877," he again touches on the old baronetcy.

91. More Advice to "Lazybones." 8th May 1877.

"Next Saturday I expect to be at Trieste, whence this letter will start. The *Times* has probably told you the story of my last adventure, and this will probably have explained to you why yours of March 8th has remained so long unanswered. That document informed me that 'Lazybones' was going to make himself useful. I hope he has done so. If not, he can learn all about his grandfather from papers published by the late Admiral Burton, and I do not think that Miss Eruli would object to letting him have copies. Of course, don't speak about the baronetcy. That failing, all he has to do is to put the matter (after making an agreement) into the hands of a professional man, who will visit Shap (Westmoreland) and Galway, and who will find no difficulty in establishing direct descent. Please write to me again. I shall be heard of in Trieste for some time. Many thanks for the good news about poor Edward. Give my best love to the boys, and salute 'Lazybones' according to his merits."

In due time Burton arrived at Cairo, and the curious expedition set forth for wild, mysterious Midian. He himself knew nothing of engineering, but he had the services of a practical engineer—one M. Marie; and some artists, and a number of Egyptian officers and Soudanese soldiers accompanied the expedition. The party included neither metallurgist nor practical prospector[1]

[1] Preface to *Midian Revisited*, xxxiv.

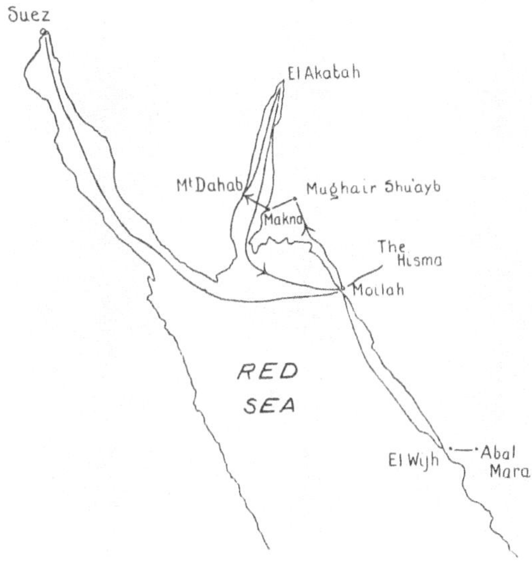

THE JOURNEY TO MIDIAN

but Burton carried a divining rod, and seems really to have believed that it would be a help. The expenses, it was ascertained, would amount to one thousand nine hundred and seventy-one pounds twelve shillings and sixpence—no very extravagant sum for purchasing all the wealth of Ophir.

At Zagazig they were joined by the venerable wag and trickster, Haji Wali, and having reached Suez they embarked on the gunboat, the "Mukhbir," for Moilah, which they reached on December

92. Haji Wali Again.

19th. Burton landed with studied ceremony, his invariable plan when in the midst of savage or semi-civilized people. The gunboat saluted, the fort answered with a rattle and patter of musketry. All the notables drew up in line on the shore. To the left stood the civilians in tulip-coloured garb, next were the garrison, a dozen Bashi-Bazouks armed with matchlocks, then came Burton's quarry men; and lastly the escort—twenty-five men—held the place of honour on the right; and as Burton passed he was received with loud hurrahs. His first business was to hire three shaykhs and 106 camels and dromedaries with their drivers. The party was inclined to be disorderly, but Burton, with his usual skill in managing men, soon proved who was master.

Nothing if not authoritative, he always spoke in the commanding voice of a man who brooks no denial, and, as he showed plainly that acts would follow words, there was thenceforward but trifling trouble. He himself was in ecstasies. The Power of the Hills was upon him.

The exploration was divided into three journeys, and between each and the next, the expedition rested at Moilah. The first or northward had scarcely begun, indeed, they had got no further than

93. Graffiti.

Sharma, before Haji Wali found it convenient to be troubled with indigestion in so violent a form as to oblige him to return home, which he straightway did with great alacrity. His object in accompanying the expedition even thus far is not clear, but he evidently got some payment, and that the expedition was a hopeless one he must have known from the first. The old rogue lived till 3rd August 1883, but Burton never again met him.

Even in Midian, Burton was dogged by Ovid, for when he looked round at the haggard, treeless expanse he could but exclaim, quoting the *Ex Ponto*,

> "Rara neque haec felix in apertis eminet arvis
> Arbor, et in terra est altera forma maris."
>
> ["Dry land! nay call it, destitute of tree,
> Rather the blank, illimitable sea."](1)

The expedition then made for Maghair Shu'ayb, the Madiama of Ptolemy and the old capital of the land. Here they spent a "silly fortnight, searching for gold," which refused to answer even to the divining rod. They saw catacombs—the Tombs of the Kings—some of which were scrawled with graffiti, laboured perhaps by some idle Nabathaean boy in the time of Christ. They found remains of furnaces, picked up some coins, and saw undoubted evidences of ancient opulence. That was all. Thence they made for Makna, passing on their way a catacombed hill called "the Praying Place of Jethro," and a shallow basin of clay known as Moses' Well. From Makna, where they found their gunboat waiting for them, they then cruised to El Akabah, the ancient Eziongeber, in whose waters had ridden the ships of Solomon laden with the merchandise of India and Sheba. They reached Moilah again on February 13th. The second journey, which took them due East as far as the arid Hisma, lasted from February 17th to March 8th. Burton considered the third journey the most important, but as they found nothing of any consequence it is difficult to understand why. First they steamed to El Wijh, in the "Sinnar," which had taken the place of the Mukhbir, and then marched inland to the ancient mines of Abul Maru. But Burton now saw the futility of attempting to proceed

(1) *Ex Ponto* III., i., 19.

further. On April 10th they were back again at El Wijh, on the 18th at Moilah and on the 20th at Suez.

In the meantime, Mrs. Burton had left Trieste, in order to join her husband. She stayed a week at Cairo, where she met General Gordon, who listened smilingly to her anticipations respecting the result of the expedition, and then she went on to Suez. Writing to her nieces, the Misses Stisted, 23rd March 1878, she said: "I have taken a room looking across the Red Sea and desert towards Midian, and hope at last to finish my own book [*A.E.I., Arabia, Egypt and India*]. What on earth Paul is doing with Richard's *Midian*[1] God only knows. I have written and telegraphed till I am black in the face, and telegrams cost 2s. 6d. a word. At last on 20th April, while Mrs. Burton was in church, a slip of paper was put into her hand: "The 'Sinnar' is in sight."

Determined that the Khedive should have something for his money, Burton and his company had, to use Mrs. Burton's expression, "returned triumphantly," with twenty-five tons of minerals and numerous objects of archæological interest. The yield of the argentiferous and cupriferous ores, proved, alas! to be but poor. They went in search of gold, and found graffiti! But was Burton really disappointed? Hardly. In reading about every one of his expeditions in anticipation of mineral wealth, the thought forces itself upon us that it was adventure rather than gold, sulphur, diamonds and silver that he really wanted. And of the lack of that he never had reason to complain.

An exhibition of the specimens, both mineralogical and archæological, was held at the Hippodrome, and all Cairo flocked to see "La Collection," as the announcement expressed it, "rapportée par le Capitaine Burton."[2] The Khedive opened the exhibition in person, and walked round to look at the graffiti, the maps, the sketches

(1) *The Gold Mines of Midian and the Ruined Midianite Cities* (C. Kegan Paul and Co.) It appeared in 1878.

(2) *The Land of Midian Revisited*, ii., 254.

of ruins and the twenty-five tons of rock, as nobody had more right; and Burton and M. Marie the engineer accompanied him.

"Are you sure," enquired the Khedive, pointing to some of the rocks, "that this and this contain gold?"

"Midian," replied M. Marie, blandly, "is a fine mining country."

And that information was all the return his Highness got for his little outlay of one thousand nine hundred and seventy one pounds twelve shillings and sixpence.

Returned to Trieste, Burton once more settled down to his old dull life. The most interesting letter of this period that has come to our hands is one written to Sir Henry Gordon,[1] brother of Colonel, afterwards General Gordon.

94. Letter to Sir Henry Gordon, 4th July 1878.

It runs: "Dear Sir, I am truly grateful to you for your kind note of June 30th and for the obliging expressions which it contains. Your highly distinguished brother, who met my wife at Suez, has also written me a long and interesting account of Harar. As you may imagine, the subject concerns me very nearly, and the more so as I have yet hopes of revisiting that part of Africa. It is not a little curious that although I have been in communication with Colonel Gordon for years, we have never yet managed to meet. Last spring the event seemed inevitable, and yet when I reached Suez, he had steamed south. However, he writes to me regularly, scolding me a little at times, but that is no matter. I hope to be luckier next winter. I expect to leave Trieste in a few days[2] and to make Liverpool via long sea. Both Mrs. Burton and I want a medicine of rest and roast beef as opposed to rosbif. Nothing would please me more than to meet you and talk over your brother's plans. My direction is Athenæum Club, and Woolwich is not so

(1) Kindly copied for me by Miss Gordon, his daughter.

(2) They left on July 6th (1878) and touched at Venice, Brindisi, Palermo and Gibraltar.

difficult to explore as Harar was. Are we likely to meet at the British Association ? ''

Burton and his wife reached London on July 27th (1878). Presently we hear of them in Ireland, where they are the guests of Lord Talbot of Malahide, and later he lectured at various places on "Midian" and "Ogham Runes." Again Gordon tried to draw him to Africa, this time with the offer of £5,000 a year, but the answer was the same as before. Then came a great blow to Burton—the death of his beloved niece—"Minnie"—Maria Stisted. Mrs. Burton, who was staying at Brighton, wrote to Miss Georgiana Stisted a most kind, sympathetic and beautiful letter—a letter, however, which reveals her indiscreetness more clearly, perhaps, than any other that we have seen. Though writing a letter of condolence—the sincerity of which is beyond doubt—she must needs insert remarks which a moment's consideration would have told her were bound to give offence—remarks of the kind that had already, indeed, made a gulf between her and Burton's relations.

95. Death of Maria Stisted, 12th November 1878.

She says: "My poor darling Georgy, I do not know how to write or what to say to you in such poignant grief. I think this is the most terrible blow that could have happened to Maria (Lady Stisted) and you. I do not grieve for Minnie, because, as I told Dick in my letter, her pure soul has known nothing but religion and music, and is certainly in its own proper place among the angels, but I do grieve for you with all my heart. . . . It is no use to talk to you about 'Time healing the wound,' or 'resigning oneself to what is inevitable,' but I have so long studied the ways of God, that I know He has taken the angel of your house as *He always does*, that this is a crisis in your lives, there is some change about to take place, and some work or new thing you have to do in which Minnie was not to be. I can only pray for you with all my heart, as I did at communion this morning." So far, so good, but then comes: "and have masses said to create another gem upon Minnie's crown."

Yet Mrs. Burton knew that she was writing to staunch Protestants whom such a remark would make positively to writhe. Still, in spite of her indiscretions, no human being with a heart can help loving her. She then goes on : "Please know and feel that though the world looks dark, you have always a staunch friend in me. Dick feels Minnie's death fearfully. He telegraphed to me and writes every day about it. I don't think he is in a state of health to bear many shocks just now, he is so frightfully nervous. He so little expected it, he always thought it was only one of the little ailments of girls, and Maria (Lady Stisted) was over anxious ; so it has come like a sledge-hammer upon him. I feel what a poor letter this is. but my heart is full, and I do not know how to express myself. Your attached and sympathising Aunt Zoo."

Burton was just then engaged upon his work *The Land of Midian Revisited*, and he dedicated it to the memory of his "much loved niece."

96. Burton's "Six Senses."

On 2nd December 1878, Burton lectured at 38, Great Russell Street before the British National Association of Spiritualists—taking as his subject, "Spiritualism in Foreign Lands." His ideas on Spiritualism had been roughly outlined some time previous in a letter to *The Times*.[1] He said that the experience of twenty years had convinced him: (1) that perception is possible without the ordinary channels of the senses, and (2) that he had been in the presence of some force or power which he could not understand. Yet he did not believe that any spirits were subject to our calls and caprices, or that the dead could be communicated with at all. He concluded, "I must be contented to be at best a spiritualist without the spirits." The letter excited interest. The press commented on it, and street boys shouted to one another,

(1) November 1876.

MISS MARIA STISTED.

From a photograph by J. Weston & Son, Folkestone.

LISA.

From a photograph by Guiseppe Wulz, Trieste.

"Take care what you're doing! You haven't got Captain Burton's six senses." At Great Russell Street, Burton commenced by defending materialism. He could not see with Guizot that the pursuit of psychology is as elevating as that of materialism is degrading. What right, he asked, had the theologian to limit the power of the Creator. "Is not the highest honour His who from the worst can draw the best?"(1) He then quoted his letter to *The Times*, and declared that he still held the same opinions. The fact that thunder is in the air, and the presence of a cat may be known even though one cannot see, hear, taste, smell or feel thunder or the cat. He called this force—this sixth sense—zoo-electricity. He then gave an account of spiritualism, thaumaturgy, and wizardry, as practised in the East, concluding with a reference to his *Vikram and the Vampire*. "There," said he, "I have related under a facetious form of narrative many of the so-called supernaturalisms and preternaturalisms familiar to the Hindus.(2) These studies will show the terrible 'training,' the ascetic tortures, whereby men either lose their senses, or attain the highest powers of magic, that is, of commanding nature by mastering the force, whatever it may be, here called zoo-electric, which conquers and controls every modification of matter.(3) His lecture concluded with an account of a Moorish necromancer, which reminds us of the Maghrabi incident in "the Story of Judar." When Burton sat down, Mrs. Burton asked to be allowed to speak. Indeed, she never hesitated to speak upon any subject under the sun, whether she did not understand it, as was almost invariably the case, or whether she did; and she always

(1) From the then unpublished *Kasîdah*.

(2) The famous Yogis. Their blood is dried up by the scorching sun of India, they pass their time in meditation, prayer and religious abstinence, until their body is wasted, and they fancy themselves favoured with divine revelations.

(3) *The Spiritualist*. 13th December 1878.

spoke agreeably.[1] She pointed out to the spiritualists that they had no grounds to suppose that her husband was one of their number, and stated her belief that the theory of zoo-electricity would suit both spiritualists and non-spiritualists. Then, as a matter of course, she deftly introduced the "one Holy Catholic and Apostolic Church" to which it was her "glory to belong," and which this theory of Burton's "did not exactly offend." As regards the yogis and the necromancers she insisted that her husband had expressed no belief, but simply recounted what is practised in the East, and she concluded with the remark, "Captain Burton is certainly not a spiritualist." Some good-humoured comments by various speakers terminated the proceedings. It is quite certain, however, that Burton was more of a spiritualist than Mrs. Burton would allow, and of Mrs. Burton herself in this connection, we shall later have a curious story to tell.[2]

During the rest of her holiday Mrs. Burton's thoughts ran chiefly on philanthropic work, and she arranged gatherings at country houses in support of the society for the Prevention of Cruelty to Animals. These were well attended and some enthusiasm was shown, except when there happened to be a meet of the fox hounds in the district, or when rabbit coursing was going on.

97. Still thinking of Midian. April—December 1879.

The Burtons remained in London until after the publication of Mrs. Burton's book "*A.E.I.*,"[3] and then Burton set out alone on a tour through Germany. Mrs. Burton, who was to meet him at Trieste, left London 27th April; and then followed a chapter of accidents. First she fell with influenza, and next, at Paris, when descending the stairs, which had been waxed, she "took one header from the top to the bottom," and so

(1) In short, she had considerable natural gifts, which were never properly cultivated.

(2) See Chapter xxxviii.

(3) Arabia, Egypt, India.

VIEW FROM BURTON'S HOUSE AT TRIESTE.

damaged herself that she had to be removed in a *coupé lit*.[1] She reached Trieste after "an agonizing sixty hours" and was seriously ill for several weeks. All the while, Burton whose purse, like that of one of his favourite poets, Catullus, was "full of cobwebs," had been turning his thoughts to Midian again. He still asseverated that it was a land of gold, and he believed that if he could get to Egypt the rest would be easy. Says Mrs. Burton, writing to Miss Stisted, 12th December 1879: "Darling Dick started on Friday 5th, a week ago, in high spirits. My position is singular, no child, no relative, and all new servants." She then speaks of her Christmas book, which had just gone to the publishers. She says, "It is for boys from 12 to 16, culled from ten volumes: Dick's three books on Sind, his *Goa, Falconry, Vikram, Bayonet and Sword Exercise*, and my *A.E.I.*" and she was in hopes it would revive her husband's earliest works, which by that time were forgotten. The fate of this work was a melancholy one, for the publisher to whom the manuscript was entrusted went bankrupt, and no more was ever heard of it.[2] Burton's hope that he would be able to lead another expedition to Midian was not realised. Ismail was no longer Khedive, and Tewfik, his successor, who regarded the idea as chimerical, declined to be bound by any promise of his father's. His Excellency Yacoub Artin Pasha[3] and others of Burton's Egyptian friends expressed sympathy and tried to expedite matters, but nothing could be done. To make matters worse, Burton when passing through Alexandria was attacked by thieves, who hit him on the head from behind. He defended himself stoutly, and got away, covered however, with bruises and blood.

(1) Letter to Miss Stisted.

(2) She says, I left my Indian Christmas Book with Mr. Bogue on 7th July 1882, and never saw it after.

(3) Burton dedicated to Yacoub Pasha Vol. x. of his *Arabian Nights*. They had then been friends for 12 years.

END OF VOL. I.

Made in United States
Cleveland, OH
26 September 2026

44826304R00177